ECLIPSE '99
Capture it on Film

Cover Image

The solar corona on 26 February 1998 (a detailed description is given in Chapter 7).

ECLIPSE '99

Capture it on Film

H J P Arnold

Institute of Physics Publishing
Bristol and Philadelphia

British Library Cataloguing-in-Publication Data

A catalogue record for this book is available from the British Library.

ISBN 0 7503 0619 X

Library of Congress Cataloging-in-Publication Data are available

Published by Institute of Physics Publishing, wholly owned by The Institute of Physics, London

Institute of Physics Publishing, Dirac House, Temple Back, Bristol BS1 6BE, UK

US Office: Institute of Physics Publishing, The Public Ledger Building, Suite 1035, Independence Square, Philadelphia, PA 19106, USA

Typeset in TEX using the IOP Bookmaker macros
Printed in the UK by J W Arrowsmith Ltd, Bristol

To the memory of Donald F Trombino 1940–1998

Sol Invictus

Contents

Acknowledgments

I am extremely grateful for the assistance in preparing this guide that I received from a number of friends and business colleagues, including Judy Watson and Stephen Griffiths of Kodak Limited; Ninian Boyle of Venturescope; David Lawrence of Broadhurst Clarkson and Fuller; Reg Parker of R J Parker & Son; Jerry Pellett; Richard Bligh; David Hammersley; and Akira Fujii (who allowed me to use his beautifully executed time-lapse picture that appears on page 46). In addition, I owe a debt of thanks to two long standing and knowledgeable friends—Leonard Dougherty (past Director of the British Astronomical Association's Solar Section) and Geoffrey Crawley (Technical Editor of the *British Journal of Photography*)—for discussions of various aspects of the themes dealt with in the book, for reading it and for making suggestions for its improvement.

All of the eclipse timings (expressed as British Summer Time) quoted in this primer are taken from *NASA Reference Publication 1398: Total Solar Eclipse of 1999 August 11* prepared by Fred Espenak and Jay Anderson. The maps are also based, with thanks, on those appearing in this indispensable guide.

All images, save in two cases as indicated in the captions, were taken by the author.

H J P Arnold
Havant, Hampshire

Introduction

Our Sun—the Star of the Eclipse

On a clear night, and away from city lights, we can see many stars in the dark sky. As the Sun rises the stars fade from view and the strength of daylight is so great compared with night time that it does take a little effort to remember that one star is still visible after all the others have gone—for our Sun is a star, much like many of the others. Often you will hear astronomers referring to the Sun as being a 'very average' star but we need to remember that, average or not, without the Sun there would be no life on Earth—and, quite possibly, no Earth at all.

The Sun is not a solid body but a ball of gas which shines as a result of a thermonuclear reaction in which the nuclei of hydrogen atoms (which make up over 70% of its volume) are combining to form the nuclei of helium. Although this reaction causes energy to be released it is at the expense of losing some mass—in fact, around 4000 tonnes per second. This may sound an enormous amount and a little worrying but there is nothing to fear because the size of the Sun is such that it has been 'burning' for around 5000 million years but is only about half way through its life!

Its equatorial diameter is almost 1.4 million km (865 000 miles)—well over one hundred times greater than that of Earth and something like one million Earths could be packed into its volume. The Earth orbits around the Sun at an average or mean distance of almost 150 million km (just under 93 million miles) which fortuitously for us is just about right for life to have evolved—much closer to the Sun would have been too hot, much further away would have been too cold.

The bright yellow surface of the Sun as seen from the Earth is called the *photosphere*, where the temperatures reach 5500 °C and where sunspots, which are basically magnetic phenomena, are located. The umbra and penumbra of sunspots are only a few hundreds of degrees cooler than the surrounding photosphere but that is enough for them to appear much darker. Above the photosphere is the *chromosphere*, a region of more rarefied gas, pinkish-red in colour, up to about 10 000 km deep with temperatures beginning at 8000 °C. This is the region where *prominences* are seen—masses of red hydrogen gas lifting off from the limb or edge of the Sun, either quietly or violently, and which can reach deep into space. Always violent are flares—sudden releases of energy that can send shock waves out into the corona and beyond. They are usually associated with the active regions around sunspots, are normally short lived, and typically cannot be seen in ordinary 'white light' but only with special instruments or filters. Above the chromosphere is the solar *corona*, or outer atmosphere—an extensive area of even more rarefied gas at temperatures of an estimated 1000 000 °C.

Usually all we see of the Sun with the naked eye is the disc of the photosphere when it is greatly reduced in brightness by clouds, in the early morning or in the very late evening just before it sets. Occasionally very large sunspots can be seen with the naked eye at these times. However, during a solar eclipse, the brilliant light of the photosphere is blocked out by the Moon (when completely blocked out this is known as the period of *totality*) and during the first few seconds and very last seconds of this period the thin, pink-red rim of the chromosphere together with any prominences which exist at that time are usually seen. As totality progresses, more and more of the corona appears as a misty white cloud extending out for hundreds of thousands of kilometres into space.

The likelihood of seeing sunspots through special filters in the hour or so either side of totality when the Sun's disc is only partly blocked or occluded by the Moon, as well as the form of the corona, depends to a considerable degree on where the Sun is in its repeating, approximately eleven-year, cycle of activity which passes from minimum to maximum and back again. On 11 August 1999, the Sun will be close to the period of maximum activity so that considerable numbers of spots can be expected. During periods of the solar minimum, the corona normally displays numerous elegant polar plumes extending far into space but at a time of maximum or near maximum, as in 1999, it is likely to be less well structured with enormous streamers stretching randomly far into space.

For many years solar eclipses were eagerly awaited by scientists because it was only then that many features of the Sun could be studied unhindered by the overpowering brightness of the photosphere. The development over time of special instruments which could be used to create artificial eclipses as seen from the Earth's surface, but more especially the ability to launch instrument-laden spacecraft beyond the obscuring layers of our atmosphere out to locations of eternal sunlight in the solar system, have greatly reduced the importance of eclipses to professional solar physicists. Some important experiments are certainly still conducted but solar eclipses are chiefly memorable now for the fact that they provide many thousands of inveterate eclipse chasers world-wide, as well as ordinary folk who happen to find themselves in a path of totality, with a chance of seeing perhaps the most awe-inspiring sight that Nature can offer.

Chapter 1

What is an Eclipse?

As the Earth orbits the Sun and the Moon orbits Earth, there is a regular cycle of events in which the Moon is opposite the Sun as seen from Earth (when a full Moon occurs) and then moves to a position between the Sun and the Earth (the time of new Moon). Sometimes the alignment of the three bodies is such that, at a time of full Moon, the Earth throws a shadow out into space which totally covers the Moon. Diagram A is a representation of this event although Sun, Earth and Moon are not to scale. This is a *lunar eclipse*. The Earth is very much bigger than the Moon (its diameter is more than 3.5 times greater) and casts a large shadow which at the distance of the Moon's orbit measures around 9200 km in diameter. The Moon on such an occasion can pass totally into the Earth's shadow or umbra (a total lunar eclipse); partly into it and partly into the penumbra (a partial lunar eclipse); or simply move through the outer shadow or penumbra (a penumbral eclipse), the last being scarcely noticeable unless a close study is made. During a lunar eclipse, the event will be witnessed anywhere on Earth where the full Moon is up and the sky is clear. The Moon does not merge into the dark sky on such an occasion because sunlight is refracted by the Earth's atmosphere onto its surface, which can appear a bright orange-red in hue when the Earth is relatively clear of cloud.

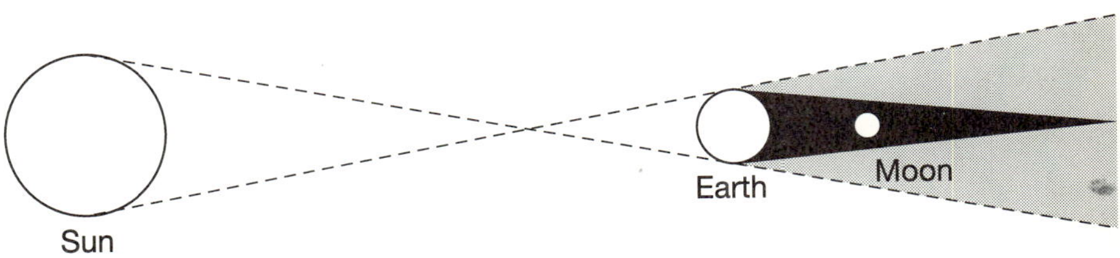

A

A *solar eclipse* occurs when the Moon comes between the Sun and Earth and blocks out the light from the Sun but the situation is a good deal more complex than that of a lunar eclipse. The Sun is enormously bigger than the Moon but by pure chance the size of the Moon as seen from Earth is just about the same as that of the Sun as seen from Earth—roughly a half of one degree. A question immediately suggests itself. When the Moon is on the other side of Earth from the Sun, we have a full Moon but since the Earth is between the Sun and the Moon why is there not a lunar eclipse, too? Similarly, if at every new Moon, the Moon comes between the Sun and Earth and they are about the same size as we see them, why is there not a solar eclipse on every occasion? There are several reasons.

The Moon's orbit about Earth is tilted or inclined by about 5° to the plane of the Earth's orbit about the Sun. This means that a precise alignment of the three bodies is nowhere near as frequent as would be the case if both Moon and Earth orbited in the same plane. Moreover, the Moon's orbit about Earth is not a circle but an ellipse so that at different times in its orbit its size as seen from Earth will appear to diminish and then to increase. Nor is that all: the Earth's orbit about the Sun is an ellipse, too, so that exactly the same thing happens in the case of the Sun. As seen from Earth, the size of its disc appears to grow or diminish depending on where the Earth is in its orbit. These factors provide the answer to the questions posed above because a series of variables need to coincide before either lunar or solar eclipses can occur.

When the alignment between the three bodies is sufficiently critical for a solar eclipse to take place, it can vary in its nature as shown in the diagrams. Diagram B portrays what happens during a *total solar eclipse*—when all of the variables come together and the Moon completely covers the face of the Sun. An umbra or deep shadow will reach the surface of the Earth where the Sun will be totally eclipsed. In the penumbral shadow thrown by the Moon, people will see a partial eclipse. A *partial solar eclipse*—and only a partial eclipse—will also occur when the Moon, even though as big as the Sun or even bigger, passes across the solar disc but does not completely cover it as seen from any point on the Earth's surface (diagram C). Then, there are occasions when the size of the Moon as seen from Earth is smaller than that of the Sun. This can be caused by the Moon being at a far point in its orbit about the Earth (and therefore appearing smaller) and/or by the latter being closer to the Sun in its elliptical orbit (and therefore the solar disc appearing bigger as seen from Earth). As the Moon moves centrally or almost centrally across the Sun, an outer ring of the bright surface of the Sun remains uncovered and the shadow thrown by the Moon does not quite reach the Earth surface. Diagram D portrays the situation. In this case, the greatest extent of the eclipse results in those on Earth in the area immediately under the Moon's umbra at this time seeing a ring of the bright solar surface of varying extent remaining in what is called an *annular solar eclipse* (from the Latin *annulus* for ring). Away from that immediate area, a partial solar eclipse will once again be seen.

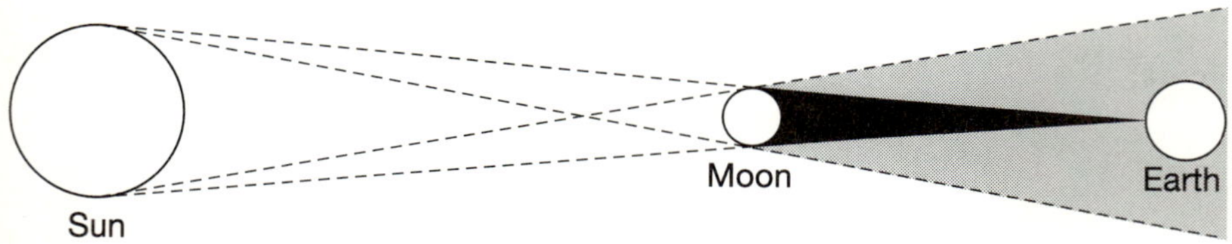

B

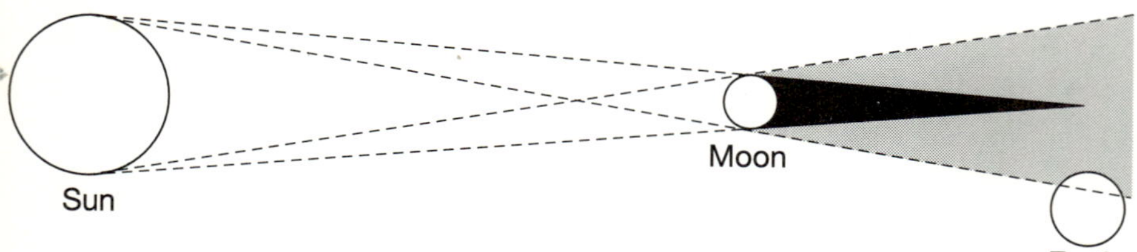

C

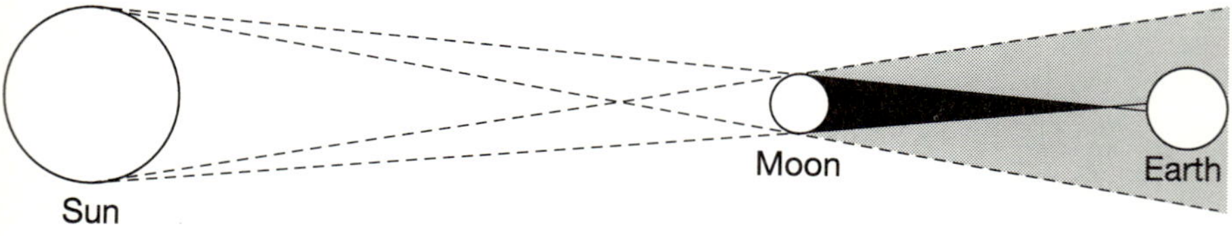

D

Reference was made above to the large size of the Earth's shadow thrown into space. Because the Moon is smaller, the shadow which it throws towards Earth when it comes between Earth and Sun is smaller—about 270 km in width at its largest. (Figure 1 dramatically records the Moon's shadow over northern South America during the eclipse on 26 February 1998.) This means that it is not a case during a solar eclipse of the eclipse being seen from anywhere on Earth where the Sun is up and the sky is clear. As the Moon moves eastwards in its orbit and occludes the Sun its shadow will touch down on the Earth's surface and will then speed eastwards until the relative positions of Sun and Moon change sufficiently for the shadow to lift off the Earth's surface and move into space. The relative narrowness of the track of totality (which frequently passes over ocean or largely uninhabited areas) means that, even given good weather, some total eclipses are seen by a relatively small number of people. However, partial phases of a total solar eclipse are seen over a far greater area and are thus not uncommon to a large number of people. Map (a) in Chapter 2, recording the extent of the 11 August 1999 eclipse, starkly compares the difference in area covered by the track of totality with that of partial eclipse visibility.

Solar eclipses have fascinated humans for thousands of years—in early history for mainly religious or mystical reasons and then increasingly for scientific reasons. A large body of knowledge (and statistics) has been built up and for those readers interested in pursuing the theme some references are given in appendices C and D. Suffice it to conclude this brief outline with a few intriguing facts.

- If the Moon were significantly further away from the Earth (or smaller) than it is now, the Earth closer to the Sun (or the Sun considerably larger) than now, we would never see a total solar eclipse on Earth.
- A total solar eclipse cannot last for longer than 7 m 31 s (except for observers aboard a fast aircraft such as Concorde flying eastwards with the lunar shadow), which means that the 1999 eclipse, with a duration of just over two minutes in SW England, will be on the shortish side in relative terms.
- The maximum number of solar and lunar eclipses of any kind that there can be in any one year is seven and the minimum number four. (In the year 2000, for example, there will be no less than four partial eclipses of the Sun but no total solar eclipse. There will be two total eclipses of the Moon, one of which will be seen from the UK.)
- Finally, one calculation is that any one spot on the Earth's surface will experience a total solar eclipse on average once in four hundred years.

Which means—enjoy 11 August 1999!

Chapter 2

11 August 1999

Central Europe has not experienced a total eclipse for almost forty years and it is thought that more people will witness the 1999 eclipse than at any time in human history. The Moon's umbral shadow first touches down in the North Atlantic some 300 km south of Nova Scotia at a few seconds before 10.31 BST when the maximum length of totality is 47 seconds and the track is 49 km wide. At 11.10 BST it reaches the Scilly Isles where the Sun is 45° above the eastern horizon, the centre line totality lasts for two minutes and the path width has grown to 103 km as the shadow moves at 3276 kph. One minute later it reaches the coast of Cornwall. However, Plymouth in Devon is the largest city on the track: although north of the centre line, it will enjoy a total eclipse lasting just under 1 m 40 s. Less than 5 minutes after the Cornish landfall, the umbra heads out into the Channel and four minutes later the centre line touches down in northern France. The northern limit of totality then moves across southern Belgium and Luxembourg before heading into Germany. Stuttgart is near the centre line and will enjoy 2 m 17 s of total eclipse while Munich, quite some way to the south of the centre line, will nonetheless experience more than two minutes of totality. At this time (just after 11.35 BST), the Sun's altitude in the sky will be 55°, the path width 109 km and the ground velocity 2664 kph. The umbra then enters Austria and Hungary and just touches Yugoslavia before crossing into Rumania. It is in Rumania that the 'instant of greatest eclipse' occurs at just after 12.03 BST: totality will last for 2 m 23 s with the Sun at an altitude of 59°, a path width of 112 km and a shadow velocity of 2448 kph. The centre line then proceeds over the far north-east of Bulgaria before heading across the Black Sea and into Turkey where the centre line duration starts a gradual decrease. A little after 12.45 BST the umbra moves into Iraq after touching Syria briefly and then passes into Iran. By 13.22 BST it reaches Pakistan where Karachi experiences a total eclipse lasting 1 m 13 s with the Sun 22° above the western horizon. At this time, the path width is down to 85 km but the umbra velocity has increased to 7200 kph. By 13.28 BST the shadow reaches India, the last nation in its path. The shadow's velocity increases rapidly with centre line totality lasting less than one minute and with the Sun's altitude down to 7° in the west. The shadow finally sweeps out into the Bay of Bengal where it lifts from the face of the Earth, having travelled along a path around 14 000 km long over the course of 3 hours 7 minutes in the last total solar eclipse of this millennium.

Map (a) shows the track of the solar shadow on a projection of the Earth's surface. Outside of that track, large areas will witness a partial eclipse—the percentage values marked on the map being the extent of obscuration of the Sun's disc. Event timings (expressed as British Summer Time) are given on the vertical bars. Map (b) sharply narrows the area covered down to the UK and France and shows the track of totality crossing Cornwall and Devon. Notice how the duration of totality shortens as one moves away from the centre line whether in a southerly or northerly direction. All areas outside the umbral path will experience a partial solar eclipse although even in the far north of Scotland the phase will be extensive—only falling to below 70% in the Shetlands. Appendix A gives detailed timings of the eclipse and other information for towns within the umbral path and appendix B gives these for a selection of towns experiencing partial phases of the eclipse. Map (c) illustrates the predicted phase of eclipse for a selection of towns in appendix B. An explanation of the various headings is given beneath the tables. Notice in the case of Newton Abbot the point about totality being shorter the further away one gets from the centre line—30 seconds is all the town gets! However, there are compensations for a location like this, as will be explained a little later.

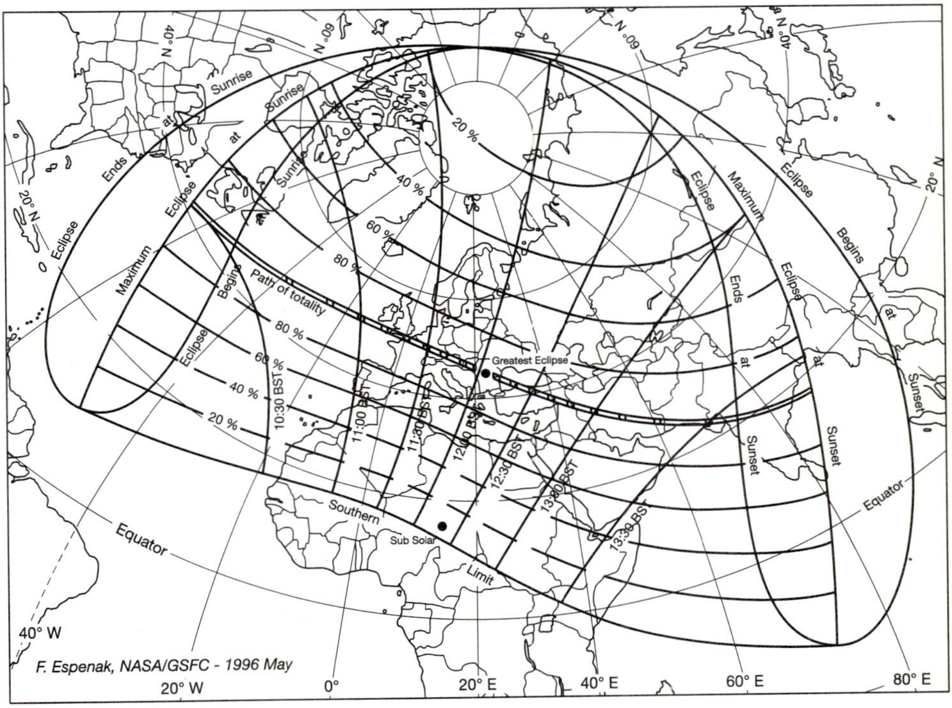

(a)

The moment when the Moon first touches the solar disc is called *first contact*. Amongst keen astronomers and eclipse chasers (with all filters and other viewing aids firmly in position) there is always a rivalry to be the first to spot the first 'nick' in the solar disc and to shout 'first contact!'. Along the south of the country this will take place at somewhat after one o'clock if we liken the solar disc to a clock face. Locations further to the north will see first contact occurring slightly lower down on the top right of the solar disc. The Moon will move steadily down and across the solar disc over a period of more than an hour before the magical moment of *second contact*—totality—arrives.

During most of the period between first and second contact, casual observers will notice little difference. Daylight alters little and nothing much happens: dare I say that for many—even some eclipse chasers—this is regarded as a period for patience and hopeful expectation, particularly as regards the weather. It is in the last fifteen minutes or so before second contact that noticeable changes occur. The light level begins to drop and the temperature does, too. Birds will start to roost and other animals may well act as though night were coming on. In an urban area, street lights activated by light sensors will come on. The excitement grows and in the last few minutes the sky darkens and takes on the appearance of late twilight although the hue of the light (to my eyes) is a strange bluish grey. People look to the west where the sky is darker still and then the Moon's shadow comes rushing in—and second contact! In the last few seconds before the Moon's disc temporarily obscures the Sun, the last remnants of the photosphere shine through the valleys and canyons of the Moon's leading (western) limb or edge—the bright, white areas of light being called Baily's Beads after the astronomer who described them in detail in the 19th century. It is at this point that all filters can be removed. A final, stark bead and the first tentative appearance of the solar corona creates the effect known aptly as the Diamond Ring. Both Baily's Beads and the Diamond Ring—assuming they are seen—last for only a second or two and are followed by the appearance of any reddish prominences—fountains of gas and plasma erupting from the edge of the Sun—which may be occurring at that time and which might well be seen together with the thin red rim of the lower atmosphere of the Sun known as the chromosphere. They will be slightly longer lived but then the beautiful pearly white colour of the corona—the Sun's outer atmosphere—will come to dominate more and more until it might well stretch out for several solar radii. Totality is now in full swing and planets, bright stars and any comets which happen to be passing will come into view. In August 1999, brilliant Venus will be close to the Sun on one side with Mercury on the other, while Saturn and Jupiter will be far away to the west. Regulus in Leo will be alongside the Sun and above Venus, with Castor and Pollux in Gemini above and to the west, bright Sirius lower towards the south west and Betelgeuse and Rigel in Orion still further to the south west. (Figure 2 is a computer simulation of the sky seen at mid-totality from Redruth in Cornwall.) As luck has it, the eclipse will be occurring very close to the maximum of the Perseid meteor shower so some 'shooting stars' may also be seen.

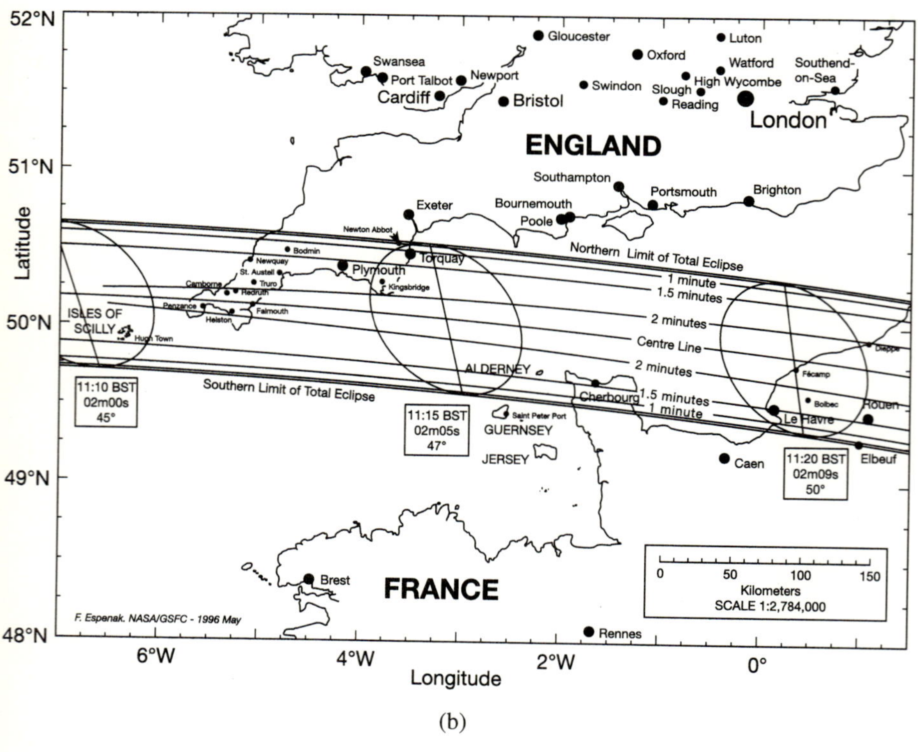

(b)

But the two minutes or less of totality will soon be over. *Third contact* occurs. The Diamond Ring and Baily's Beads herald the return of the photosphere, this time at the trailing edge or limb of the Moon—at which time filters and viewing aids must be immediately put on again. For most people the shouts and gasps of amazement and delight at the celestial beauties witnessed during totality are replaced by excited chatter and discussion as impressions are compared. For many people the eclipse is basically over—only those made of the 'right stuff' are interested in such mundane matters as looking at or recording the partial phases of the Sun as the Moon begins to move off the disc—and now is the time that a celebratory drink is consumed even by (or perhaps it should it be said, particularly by) the inveterate eclipse chasers. There can however be variations: after the eclipse seen from India in 1995, fireworks were the order of the day. It takes another hour for the Moon to clear the disc of the Sun and for what is called *fourth contact* to occur.

Reference was made earlier to the shortness of totality at Newton Abbott in Devon but there are compensations. It is quite true that the length of totality is the greatest at the centre line and that there the corona is seen to its greatest extent. But Baily's Beads, the chromosphere and prominences are seen far better and for a longer time along the 'eclipse edge'. In fact some eclipse expeditions deliberately choose to locate themselves as closely as possible to the edge of the umbra. I joined one such group for the eclipse of 3 November 1994 which was observed from the Atacama Desert in Chile. GPS (Global Positioning Satellite) sets were used to locate our position where totality lasted for a little under a minute compared with a centre line totality of somewhat under 3 minutes. It is quite difficult to predict precisely where the edge of the umbra will occur in any locality, which means that the inhabitants of such small villages as Port Isaac and Portgaverne on the Cornish coast and Treburley further inland, together with Bickington, Newton Abbott and Teignmouth across in Devon, should have an exciting if short-lived eclipse as the Moon just about covers the solar disc—although Teignmouth on the north side of the Bovey Estuary may be just a little too far to the side of the track. (Computer programmes can differ on the data taken as the basis for calculations and this is another reason for some uncertainty—for example, in the timing of eclipse events.)

Inevitably cold reality must enter the picture and it needs to be stressed that all of the above presupposes that the Sun can be seen at the critical period on the morning of 11 August 1999. NASA's 'Reference Publication' on the 1999 total solar eclipse does not mince its words as to the weather possibilities at home: '. . . Climate statistics suggest that England is the least suitable land location from which to view the eclipse except for parts of India and Pakistan where monsoon cloudiness dominates . . . Surface observations show an average of 5.8 to 6.5 days of the month with scattered cloud and good visibility, slightly lower than much of the rest of Western Europe, but only a third of the cloud-free days available in Bulgaria and Rumania, and a quarter to a fifth the number of days in the best locations in Turkey and Iran.' Given the present international situation, probably few even of the keenest would contemplate Iran but the above explains why many of the specialist astronomical eclipse parties are journeying to Rumania, Bulgaria and Turkey. I know from personal experience that the objective is to give the maximum chance of seeing the eclipse by selecting a location with the best odds. However, once the major decision has been taken, solar eclipses teach one to adopt a fatalistic attitude. Whatever will be, will be. Of course, the odds can be improved a little in certain circumstances. Given conditions of extensive but patchy cloud in Cornwall and Devon on 11 August 1999, there could be every good reason to plan to take to the sea, provided there were no intention of conducting serious photography (see later chapters). In some parts of the world there might also be a temptation to add that the availability of a motor vehicle could be advantageous in such weather conditions but this is highly unlikely to be the case on the highways and byways of Cornwall and Devon on the morning of 11 August 1999 where police road closures might be the order of the day!

The important thing is that statistics can be a valuable pointer but may not necessarily reflect what happens on the day. There may be a 4 in 10 chance of seeing the eclipse in Cornwall and a greater than 8 in 10 chance of seeing it in south-eastern Turkey. But 11 August 1999 may fall on one of the 4 good days in Cornwall but on one of the cloudy two in Turkey! Whether it is your wish to go to Cornwall and Devon, or perhaps your purse dictates that location, lay plans to do all that you want to do assuming good weather: the skies could be clear.

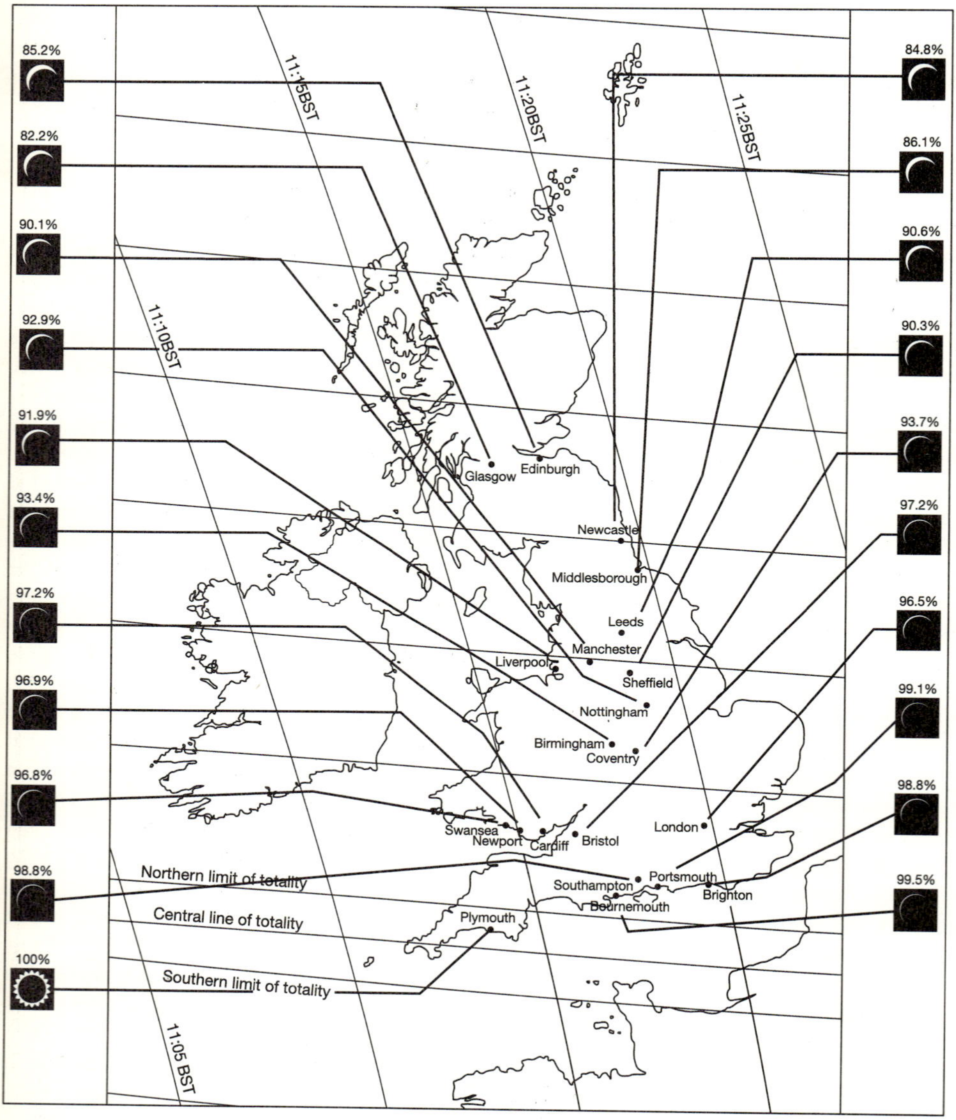

(c)

Chapter 3

Safety

A solar eclipse is a time of great excitement but for the uninitiated it is also one of danger which needs to be described without exaggeration. Even those readers with no knowledge whatsoever of astronomy will be aware from past experience of what happens if we look up to a sky filled with patchwork clouds and then the Sun suddenly comes out from behind a cloud: we are dazzled for a short while and the eyes may take some time to recover. Then again, we may recall occasions that we have experienced or that we have read about when the Sun, shining through a lens or glass of some kind, has been focused on combustible material and the heat generated has led to a fire. There is thus an element of danger when an awe-inspiring celestial event results in many inexperienced people turning their eyes to the Sun. It needs to be emphasized here that this danger exists only during the partial phases of the solar eclipse on either side of totality. The Sun can be looked at during totality with complete safety, not to say comfort, to the eyes. But in a sense there are no half measures about a solar eclipse: only a fraction of one per cent of the solar photosphere may be showing just outside of totality close to the edge of the track, but that is enough to require protection of the eyes in the manner outlined here.

The danger from the Sun outside of totality is that, while we might be viewing it through a medium which attenuates the visible light and therefore makes looking at it perfectly comfortable, our eyes could be exposed to infrared and ultraviolet radiation capable of damaging both the lens of the eye and more especially the retina which is the 'screen' of the eye. Photochemical and thermal damage of the retina would not be apparent immediately because the retina has no pain receptors, so no warning signals would be received until impaired vision developed later. The danger therefore is largely (though not entirely) an insidious one, which constitutes a greater threat.

Often astronomical books about the Sun and solar eclipses are quite categorical in stating that there is only one completely safe way of looking at the Sun and that is by projection. In this method, a pair of binoculars or a telescope are pointed toward the Sun and its image is projected onto a light card or screen at a distance where the image is in as sharp focus as possible. The result is improved if a large cover of some kind is arranged around the instrument so that the area where the image is projected is darkened as much as possible. The projected image can be studied with complete safety to the eyes and this is the method employed by some keen amateur astronomers who, for example, endeavour to keep a count of sunspot numbers on the Sun.

In my view this advice is proffered with the intention of counteracting the lack of good sense on the part of some people when alternative and perfectly safe means of looking at the Sun are available. These means are both far more convenient and superior in terms of the quality of the image of the Sun seen but they do demand that the users follow one or two simple instructions. It is for individuals to decide what is done and for the time honoured guidance of 'if in doubt, don't' to be applied.

Nonetheless, a number of perfectly safe filters for looking at the Sun on either side of totality do exist and at least one type will be readily available for naked-eye viewing of the Sun as well as for filtering instruments such as binoculars, cameras lenses and telescopes during the eclipse in August 1999. This is a form of aluminized Mylar—a plastic product—which comes in thin sheets which are doubled and when used to view the Sun absorb 99.999% of its radiation yielding a neutral density (ND) of 5, which is perfectly safe. This filter material is marketed under various trade names—Solar Skreen from Roger Tuthill Inc. in the USA being one of the longest established—and is offered in the form of ready-made, cardboard spectacle frames, square mounts for cameras and telescopes, as well as sheets which can be cut to a desired size. The only drawback in terms of the image is that the Sun appears as light blue in colour. So far as safety is concerned, the two thicknesses of Mylar are certainly safe for the eyes but the material does need to be treated reasonably carefully and to be checked for pin-holes or creasing. If these are present, the filter or material should be discarded immediately

As popular interest in the eclipse grows, there is no doubt that it is the various forms of Mylar viewing aids that will prove most popular. Indeed, the spectacles at a price of about £2.00 a pair were available in the UK in the summer of 1998! Be that as it may, there are other safe filters which individuals may prefer. Most serious amateur solar astronomers use glass filters which are coated with a combination of metal alloys and which again absorb 99.999% of solar radiation (save in specialist 'faster' versions intended solely for photography and not viewing). These have the advantage of showing the solar disc in its normal yellow. Further reference is made to these in Chapter 5, devoted to the keen photographer, but their relative high cost would make them unappealing to most hopeful viewers of next August's eclipse. Another safe filter for those with access to it is welder's glass with a shade value or number of 14.

There is yet another safe filter, which keen amateur photographers might well wish to make for themselves. A conventional black and white film (120 size would be best for ease of viewing) is unrolled fully in daylight so that it is totally exposed. The film is then processed normally so that is appears a dense black as a result of the metallic silver contained in the emulsion. Two pieces of the film of a convenient size are then cut off, the more vulnerable emulsion sides are faced inwards in the sandwich for added protection, and the pieces taped together. This forms a safe filter with an ND value of 6 which yields a yellow Sun. One drawback is that it cannot be used for critical photography because the particles of metallic silver scatter light and produce a slightly less sharp image. It must be noted that the growing number of so-called chromogenic black and white films should not be used because they are processed in the same way as colour films and all of the metallic silver is removed during processing thus leaving only dyes, which transmit harmful radiation.

The list of other unsafe filters includes smoked glass which has a long, if undeserved, tradition as an eclipse aid. The deposition of carbon on the glass is uneven, easily removed and, although visible light may be attenuated sufficiently to provide comfortable viewing, infrared and ultraviolet radiation is not attenuated to any significant extent. Sun glasses, crossed polarizers, and processed colour film (unlike conventional black and white film) are also unsafe. Black and white film of any kind containing negatives is unsafe as are conventional photographic neutral density filters unless they are used in the manner described in Chapter 5. More recently, some CDs and CD-ROMS have reportedly been used as solar viewing filters but these vary in composition (to say nothing of the danger of inadvertently looking at the Sun through the hole in the middle) so all should be regarded as unsafe. And it is well to remember that outside of totality even the safe filters already referred to must be fitted to the front of any viewing instrument securely—best to get a tailor-made mount—and that the cardboard frame Mylar spectacles should not be played around with at that time. In addition, youngsters (particularly those of an adventurous disposition) need firm guidance and watching outside of totality. It may be apocryphal, but there is the eclipse story of the person who asked when viewing the partial phases through a safe plastic filter 'I can't see the Sun very clearly: can I prick a hole in the filter to let more light in?' This is the sort of thinking that leads understandably to that blanket advice to project the image of the Sun and never to view directly.

But to be frank, I have always believed that view to be unduly restrictive. Appendix E contains a list of dealers and distributors who can be relied upon for sound advice and guidance on purchases—there is now an EEC directive affecting the certification of viewing aids if this will make you feel even more reassured—and further information on safety, as well as all matters concerning the eclipse, may be obtained from the books, institutions and web pages listed in appendices C and D.

An eclipse is a magnificent sight that should be enjoyed by the greatest number of people possible—and it takes only a relatively small amount of thought, preparation and expenditure to enjoy it in complete safety.

Chapter 4

Snaps of the Sun

During a year millions of photographs are taken by people wishing simply to record the events of a holiday at home or abroad—or of some other memorable occasion. They are not interested in photography as a hobby but only as a means to an end. This chapter is directed to them in the context of the solar eclipse on 11 August 1999. Innumerable, reasonably low-priced cameras are on the market but I will concentrate here on what are usually called *throwaway*, *single use* or *fun* cameras and on the simpler APS (Advanced Photo System) or compact cameras. I do not normally use such cameras but for the purposes of this chapter I have tested a number and I have been quite surprised by the results, the quality of which is mainly due to the ability of the colour print film used in such cameras to produce reasonable results despite gross overexposure and some underexposure. I have not been able to simulate a total eclipse of the Sun in these tests but a general indication of what can be achieved is revealed by making exposures during mid-late twilight on a clear evening.

Photography at the time of a solar eclipse can be divided into three general sections:

(a) Recording events before and after the eclipse—what can be called pictures of 'atmosphere' which are not directly astronomical in character.
(b) The partial phases of the eclipse.
(c) The period of totality. (The detailed events occurring in (b) and (c) are described in Chapters 1 and 2.)

(a) The atmospheric images offer by far the best and most numerous opportunities to the snap-shotter. For those who have never been to an eclipse, the scenes and situations are best described as being something like a cross between a festival and a carnival. That being the case, all that is needed to get good photographs is to endeavour to follow sensible picture taking practice as recommended by the camera manufacturers—like keeping the camera very still, choosing camera angles carefully so that the Sun is behind or to the side, not getting too close and so on. If you are the sort who likes to construct a picture record of a holiday you will find the eclipse as suitable for such treatment, if not more so. These are some of the themes you could feature.

(i) Invariably specially designed T-shirts are produced (indeed at least one devoted to the 1999 eclipse was available here in Britain a year in advance) and sometimes baseball caps, too. The design is frequently what might be called exotic. People wearing the shirts are a good subject.
(ii) Sometimes, large advertising posters are designed on the occasion of an eclipse. These can be very impactful or funny and present no problems for a simple camera.
(iii) Not infrequently, specially branded drinks are produced to celebrate an eclipse. While normally simple cameras cannot be used close up—for example to photograph bottles with

colourful labels—pictures of people with glasses and holding a bottle or two on high will be no problem. If the eclipse has been seen without weather problems, there is a strong celebratory feeling afterwards and this is the time to get such shots, particularly if one of a growing number of British wine producers decides to sell a dedicated vintage.

(iv) At the other extreme, I have no doubt that there will be numerous traffic jams around the time of the eclipse. While subsequently causing a pained smile, they too would be part of the storyboard.

(v) If you have any serious photographers near you, they will be setting up their tripods and other equipment early. They make a good subject (but don't get in their way!) as do all those present when they don the special safety glasses or look through framed filters once the eclipse has begun.

The sort of photos that can be taken are shown in figure 3. These were not taken with simple cameras but certainly the subjects are suitable for such cameras. With the onset of the eclipse, the much more demanding astronomical aspects of the occasion begin.

(b) As the Moon begins to encroach on to the disc of the Sun—the partial phases of the eclipse—it is important to stress again the demands of safety, details of which were fully set out in Chapter 3. Everybody looking at, or endeavouring to photograph, the Sun must be wearing or be equipped with appropriate filters.

What has gone before may be described as being perfectly normal in photographic terms. What follows is certainly not since it is astronomical in nature. Recording images of the Sun as a subject in its own right is difficult and requires the correct use of filters. The Sun will be extremely small in size in any successful images obtained with simple cameras. Then again, photographing a fairly dark sky (as during totality) is not usual and presents problems. So what can be done?

Let me emphasize one disappointing fact immediately. Cameras have grown continually in sophistication and for the snapshotter who wants something a little better this has taken the form of automatic operation. Some are completely automatic in the sense that the camera measures the light and sets the controls; similarly the lens will focus on an object automatically. This renders such cameras at best doubtful for use in most forms of astro-photography including that at a solar eclipse. The Sun's size will almost certainly mean that the automatic focusing system cannot find it. If partial eclipse photography is attempted, all that the exposure sensor will see will be a black sky with the Sun scarcely registering. At totality, too, the camera will select the longest possible exposure because the conditions will be so dark. This is likely to be too long certainly for hand-held photography and could well over-expose some of the objects in the sky in any case. Things are very different if such controls can be over-ridden for manual operation, in which case for the owners of such cameras the following chapter will be more helpful. Of course, an automatic camera is perfectly fine for recording the atmospheric pictures referred to in (a) above.

So, does this mean nothing can be attempted using simple cameras during the eclipse? Surprisingly, perhaps, the answer is no—and the simpler the camera, the better. The reason is that the films loaded into throw-away cameras are quite fast, at least ISO400 and increasingly ISO800. In the case of simple APS cameras the former speed of film is available but the characteristics of the system renders this speed of film higher than its nominal rating would suggest. The shutter speed and lens aperture (opening) in such cameras, while intended for everyday photography, just happen to be right to make it possible for anybody to have a shot at photographing the partial phases of the eclipse—and also the scene during totality. Focus and exposure setting are fixed with no sophisticated mechanism being confused by what is happening.

Consider the images reproduced here. Figure 4(*a*) is a print from the cheapest form of throwaway camera. Mylar filter sheet was carefully taped over the view finder and over the lens. (This was not expensive. I cannibalized a pair of the Mylar spectacles (Chapter 3). Both viewfinder frame and the lens are small in these cameras and were covered completely by taping the Mylar carefully over them.) Also, the camera, which has no means of being securely fixed on a tripod, was held firmly on the top of a wall while the shutter release was carefully pressed. The image of the Sun is very small on the negative (and is a very light blue as a result of the use of Mylar) but when seen in the enlarged 4 × 6 inch print is certainly visible and quite distinct, such that the partial phases of a solar eclipse could be easily distinguished. Most cameras of this sort have a wide angle lens of between 30 and 35 mm in focal length which is why the Sun's image is very small. The shutter speeds are set at around 1/125th second and the focal ratio at f/11 or so, which fortuitously is acceptable in these partial phase attempts although there is a tendency to some over-exposure.

But the size of the Sun can be improved in the case of at least one of the fun or throwaway cameras. The Fujicolor QuickSnap Super 800 Tele 100 mm camera has—as its title indicates—a lens of considerably greater than normal focal length. Figure 5 is composed of two 4 × 6 inch prints from Mylar images of the Sun and the increase in the size of the solar disc is evident. Once again there is a tendency to over-exposure with the fixed settings on the camera, but it is not too extreme and demonstrates that the partial phases of an eclipse would be well recorded.

It is obviously for the individual to decide whether to try and get images of this kind. As the pictures reveal, a stark Sun will be seen in a dark, featureless sky. I can well imagine many not wishing to make the attempt and there is a risk of failure for those with no previous experience or lacking confidence. But others might well wish to try to get their own images of these stages of the eclipse. I must emphasize the need for safe filtration of both the viewfinder as well as the lens of the camera, and for the camera to be placed on a firm base.

As the last stages of the partial phase approach and during totality there is one requirement which in my view should be observed by all those using simple cameras casually or more seriously. Often throwaway cameras can be purchased with a built-in flash incorporated which cannot be switched off. My firm advice is not to buy them. The flash can do nothing for any pictures taken of the Sun as darkness approaches or of the sky during totality. Moreover, while light falls off swiftly just before totality and the length of totality is short, the eye adapts to the darkness and it is this that enables us to see the planets and stars quite quickly as the eclipse progresses. Flashes from cameras will ruin that adaptation for anybody in the general area and will therefore ruin their enjoyment (as well at that of the camera owner). Further, if there are any serious solar observers or photographers nearby, there is likely to be a heated altercation. The moral is obvious: in everybody's interest, do not use a camera with in-built flash which cannot be switched off. If you have one, and for whatever reason cannot get another for the period of totality, make sure that during the few minutes around totality you have several pieces of electrical tape or something similar to put over the flash window. One is not enough and in tests I found that three or four pieces were necessary.

(c) It is convenient to divide totality into two phases. The first is astronomical in nature but is somewhat more akin to the atmospheric theme than photographing partial phases of the eclipse or the sky at totality. As the eclipse progresses the visible area of the Sun steadily diminishes and we get the well-known effect of the leaf canopy of trees acting as myriad pin-hole cameras, with images of the Sun being projected onto the ground. Recording this phenomenon (a natural form of image projection!) is best left to the later stages of the eclipse because a relatively slender crescent of the Sun is more aesthetically pleasing than a gibbous stage (see the lower panel of figure 9). This presents no problem for photography, whether using simple cameras or more advanced APS models, other than observing any restrictions on how close you can approach to a subject. There is usually enough light left under tree canopies at this time so there should be no exposure problems.

In the last fifteen or twenty minutes before totality, you will notice birds beginning to roost. Any that are nearby in trees in numbers would be worth a frame or two as would a general scene that includes street lights being triggered by the darkening conditions. Simple cameras would need to be as firmly positioned as possible at this time—as they would if you turn to the west a second or two before totality as the Moon's shadow rushes towards your location. Simple cameras may under-expose such pictures but it is worth the risk: film is relatively cheap and you might never see another solar eclipse, so why not have a go?

Then comes totality proper. Baily's Beads, the Diamond Ring, prominences —as described in Chapter 3—will hopefully be seen before the corona appears in all its ephemeral beauty. Nor will the eclipsed Sun be on its own because very bright Venus will be seen immediately to one side as stars also come into view. What will it be possible to record with the throw-away cameras and simple compacts and APS cameras? It is difficult to predict with any certainty but figure 4(*b*) is a simulation produced by combining a picture taken at mid-twilight using a 'fun' camera with an image taken with a much more advanced camera of an eclipsed Sun showing the corona. As we have seen, the typical throw-away camera has an f/11 focal length lens with a shutter speed of about 1/125th. Those with some photographic knowledge can study the suggested exposures for different speeds of film in table 5.1 to get an idea of what this means. While the initial, immediate stages of totality will probably produce over-exposure and any extensive area of the corona will be unlikely to be recorded, at some point during totality one or two good images should result. (While there may be over-exposure in the early stages, the resulting lens flare can make for very impressive effects in a picture.) The main drawback with the typical cheap camera will be the very small size of the Sun. However, this will be increased by the corona and the recorded scene should have Venus in it too (see figure 2)—together with any residual light in the sky. Since the Sun will be very high in the sky, if you want to try to include the (very dark) foreground in the picture, you will need to use the camera in an upright position.

Serious consideration should be given to using the Kodak, Fuji and any other cameras loaded with ISO800 film during totality. The latitude of print films will probably take care of much of the over-exposure that may occur during the early period of totality but the additional speed of the film will prove very valuable in the attempt to get as much as possible of the corona and of the general sky at mid-totality. Moreover, the substantial increase of the size of the Sun using the Fuji camera which incorporates a 100 mm focal length lens could produce an impressive picture when processed and printed. This camera has one further advantage. It has two aperture settings—one of f/15 for daylight and a second of f/9.5 for use indoors or in dark conditions. It would be sensible during totality to shoot a number of frames with each, when the latter (designated for Flood light) should record a greater extent of the solar corona. Once again it is important to emphasize the importance to the end result of the camera being very securely and firmly placed on a support of some kind.

No doubt many boat trips will be on offer at the time of the eclipse. The theory is that if conditions are partly cloudy, the boat can make for any patches of sunlight at the critical time. If a boat trip is the difference between seeing and not seeing totality then its appeal is obvious. But it needs to be remembered so far as photography is concerned how many times in this chapter reference has been made to a firm support for the camera. Aboard a boat, even if the engines have been cut and on a smooth sea, there is movement and with that comes the risk of image smear. This may not matter too much since the lenses of the cameras likely to be used by many are wide angle and the Sun's disc will be very small. However, the point must be made that photography from the deck of a boat cannot improve the quality of photography and may worsen it considerably.

In reviewing and using various camera models to prepare this chapter I was reminded of what might be thought of as the basic unfairness with regard to cameras which results from the special demands that astro-photography makes. For example, Kodak Limited makes a range of APS cameras. High in the range is the Advantix 4100ix Zoom—a good quality automatic model with a fast, 30–60 mm zoom lens and many features. It costs at the time of writing one penny under £100. The lower end Advantix 1600 has a fixed focus, wide angle, f/6.6 lens and one shutter speed. It costs well under half the price of its more illustrious stablemate. But handled carefully, the 1600 should produce some reasonable images whereas it is open to doubt whether many satisfactory pictures would be obtained using the more expensive camera.

To summarize—for those using low-priced cameras, they will be attempting something more difficult that they have attempted before. But it is not impossible. My advice would be as follows.

- Enjoy everything about the eclipse by just looking most of the time.
- If you decide to give photography a go (and fortune is said to favour the brave!) remember the safety procedures with filters on both view finders and lenses, do not use flash on any account and make sure the camera has a very firm support. (Outside of totality, remember protection for the eyes generally, of course.)
- If you do attempt to get images of the partial phases, limit the number of frames—perhaps to at most six on each side of totality.
- During totality, leave most of the time simply for enjoying the spectacle. As to photography, take no more than one or two shots right at the beginning, three or four in the middle, and maybe two at the very end. *Remember that filters are removed for the duration of totality.*
- If you are likely to be keen on the photography, do not stint on film. Should you be using a throwaway camera and are attempting partial phase images, keep one camera for those and get another exclusively for totality. In this way, your Mylar filters can be left on the one and none will be needed for the other.
- When you take your films/cameras in for processing make sure you tell the staff to 'print regardless'. You may need to emphasize that. In fairness, I should point out that this sort of photography is very different from that during holidays, parties and weddings. Moreover, there is a possibility that, if the 35 mm printing equipment cannot locate a frame line because of the dark lighting conditions in a scene, it will reject the image as being blank. This is not a problem with APS films because the printers work differently.

After the tests I have conducted, I shall be very interested to see how successful the results from throwaway and very cheap cameras will be. Even if nothing is attempted with the partial phases of the eclipse, shooting pictures during totality using the ISO800 film cameras and particularly the 100 mm focal length lens camera will most definitely be worth quite a few frames. I strongly suspect that, with due care, good pictures will result.

And if those at the eclipse get just one good, memorable image of the Sun apiece it will be important because it will be *their* picture, not somebody else's. Good luck!

Chapter 5

Serious Photography

This chapter is intended for readers who are keen amateur photographers, with reasonable experience and a good selection of equipment. Astro-photography differs greatly from everyday photography, not least because most of it is conducted at night. But there is good news for the amateur contemplating covering the 11 August 1999 eclipse. Obviously, the photography will take place during a very comfortable period in the morning, not at dead of night. There are no worries about lengthy exposures of dim objects—even using the appropriate filters, there is ample light from the Sun. Light pollution is no problem, nor is the condensation which forms all too quickly on the front surface of a lens during the night hours. There will be no problem with reciprocity failure in the film over lengthy exposures—which will vary from very short to relatively short. And that bane of the hardy astro-photographer/astronomer—freezing cold temperatures—should not be encountered on an August morning, should it?

Nonetheless the eclipse will make demands on the photographer which are outside normal experience. The first requirement is to recognize the threat from Spode's Law, popularized by the inimitable Patrick Moore, which states that if something can go wrong it will and that it will go wrong in the worst possible circumstances. Examples are that it is at the end and not the beginning of a session that the photographer will discover that there is no film in the camera or that there has been a camera malfunction. Similarly, if the attempt is being made to create a time-lapse image, it will be towards the end of a lengthy period of photography, not the beginning that the camera and tripod are knocked over. Guidance on avoiding such pitfalls will be given as we proceed through this chapter. But some important advice must be offered at the outset. Thoroughly check all of the equipment that you plan to use for photography of the eclipse well before 11 August so that repairs or adjustments can be made in good time. If you have suspected for some time that the shutter mechanism or the film wind-on have not been working properly have them checked and, if necessary, repaired. If the camera depends on batteries, get new ones for the eclipse—do not chance their failing during the event. If you are purchasing a new camera to celebrate the eclipse make sure you are fully conversant will all of its features well before 11 August. And take plenty of your chosen film with you. An eclipse is a marvellous subject for your camera and you will curse yourself if you run out of film on what will hopefully be an exciting morning.

At least three areas of photography can be contemplated during the eclipse: (*a*) telephoto images of the Sun recorded during the partial phases and/or totality to obtain the maximum size of solar image on the film; (*b*) a time-lapse picture in which the entire eclipse is recorded on one frame of 35 mm film; and (*c*) what we called 'atmosphere' images in the previous chapter, taken of a wide variety of subjects. Before going into detail about all three, preceded by a description of the equipment needed, it is important to stress that, whatever programme of photography is attempted, some time should be left for simply looking at and enjoying the spectacle, particularly during totality.

Camera

The best camera for most astro-photography (and eclipse photography is no exception) is a manually operated, 35 mm single reflex camera that accepts interchangeable lenses. The latest auto-focus, auto-exposure, auto-everything marvel is not needed and can be a positive liability. The camera lenses will be set on infinity for the Sun so auto-focus is redundant. Recommended exposures for a given speed of film will be used in a programme of bracketing if my advice is followed, although to be fair an advanced camera with a spot-metering function could be used for telephoto images of a large solar disc. A motorized film wind-on is certainly no disadvantage, but a very valuable accessory if you have one, or can get one, is some form of right-angle viewer. The Sun will be high during the eclipse and with a tripod-mounted camera you will be constantly monitoring it. Looking through the normal direct viewfinder will be very tiring on the back, particularly if you suffer with it in any way. The right-angle viewer can take the form of a clip-on attachment, a waist level type which replaces the normal direct view finder or one of the more advanced right-angle viewfinders which not only replaces the standard view finder but which magnifies the resulting image. Remember that right-angle viewers laterally reverse the image (left to right) which is important when you are framing the Sun in the viewfinder for a sequence of images on one frame or for a tracking run. To the eye, the Sun on 11 August 1999 will be travelling from east to west (left to right)—but in a right-angle viewer it will be travelling from right to left and moving up, but more of that a little later.

Lenses

Probably you will not have included the Sun in your everyday photographs so a demonstration of its size using lenses of different focal length is of value. Reproduced here (figure 6) are solar images taken with 50 mm, 180 mm, 300 mm and 450 mm lenses with a fifth exposed using a 1000 mm f/11 reflex-mirror lens. The 50 mm images were exposed through neutral density filters but a metal-coated glass solar filter was used for all of the others. Bracketed exposures of 1/60, 1/125, 1/250 and 1/500th of a second at f/11 through each lens were made on Kodak Elite II ISO100 film. Both middle exposures were satisfactory according to preference—a slightly lighter or darker solar image. There is a formula for working out the size of the Sun (or Moon) using a certain focal length of lens with a 35 mm camera: divide the focal length of the lens by 110–100 is much simpler and not too inaccurate—and the resulting number is the diameter of the solar image on the 35 mm film frame expressed in millimetres. So it is very clear why the normal 50 mm lens yields such a tiny image—the Sun is 0.5 mm across on the film!

If lenses of the above focal length are available a decision has to be taken as to which to use. However, if the object is to image the Sun as large as possible then a 450 mm or somewhat longer focal length is very suitable. A 1000 mm focal length should be regarded as the absolute limit to be used because the solar corona can extend out to several solar radii and longer focal length lenses would create a risk of the corona running out of the frame. One other piece of advice on lenses: tape the focus setting on infinity because it is very easy for it to be altered inadvertently and without you noticing. The very long mirror reflex lenses (like telescopes) can be focused beyond infinity to counterbalance temperature effects but you will find that almost certainly the infinity mark will give you the optimum focus, although this can be checked in advance.

Tripod

This (or a mount of some kind) is one of the astro-photographer's most valuable pieces of equipment. Much of the photography on 11 August will be done with long lenses which are heavy or in a programme where the Sun is being tracked through the sky. A tripod is essential but not any tripod. For everyday photography a fashion has developed for lightweight tripods which are adequate for the job for which they are intended. But in photographing the Sun you will want a very firm tripod which can bear the weight of what could well be heavy lenses and yet stay rock-steady. This will usually (although not necessarily) mean a tripod which itself is heavy. Nor is that all. The head of the tripod on which the camera/lens combination is mounted must be capable of being locked rigidly when needed but also of moving smoothly horizontally and vertically as you track the Sun. This is stressed because even good tripods can have a surprising amount of 'whip' when something like the Sun is being framed in a camera view finder: the image is central but then you tighten the locking nuts and the Sun promptly rises or drops, or moves to one side. The ideal would be to purchase what is normally called a slow motion adapter—a camera mount on which knurled rods can be rotated to move the camera through two axes smoothly and by small amounts without altering the tripod head on which it is mounted. These are not very expensive—about £50 from astronomical distributors—but, unless you think you might continue with astro-photography, sufficient practice with an ordinary tripod head will probably solve most framing problems. Try a ball and socket head if you have one but it can be difficult to control, particularly when bearing a heavy camera/lens combination. It goes without saying perhaps, but the corollary of using a tripod is to add a shutter release cable to reduce camera movement and therefore possible image smear to the minimum.

Film

Choice of film risks leading to endless debate but I intend to concentrate briefly here on the two types of colour film available, and speed. Print films have improved greatly in recent years. Their quality is of a high order and their latitude (ability to produce good quality images despite extensive over-exposure in particular as well as under-exposure) is outstanding. Many deep sky astro-photographers produce magnificent images using such films. However, I have always felt unhappy about using a film type in which there is an extra step—producing a print from a negative of what might well be an object for which I have no or little colour information other than memory, which can be very misleading. This will not be the case with the Sun for obvious reasons but traditionally I have favoured positive or transparency colour films where there is only one basic form of processing with the reproduction of colour (dependent on some variables it is true) being 'locked' into the film. There is a drawback in that transparency film has far less latitude than print film but if the photography will normally include bracketing of different exposures this is of no great significance. Nonetheless, while I will be using transparency film only on 11 August, I have no doubt that many fine images will be produced using negative or print film and the choice is down to the individual.

While advice on exposing different speeds of film appears below, the use of high speed films is not necessary—unless one is contemplating hand held 'atmosphere' images or perhaps hand-held shots of the Sun with a medium telephoto during totality, the faster shutter speeds permitted by the use of faster films being an advantage (see the end of this chapter). I typically use an ISO100 transparency film for both partial phases of the eclipse as well as totality. Quality is such these days that you might wish to use an ISO200 or even 400 speed film but I do not feel it is necessary and the slower material still does have the edge when it comes to detail—for example, in recording sunspots that might be seen on the solar disc during the partial phases or prominences on the limb or edge of the Sun during totality.

There is one final piece of advice that could help. While the results of the eclipse will not resemble those from a night's photography of dim or diffuse objects in a dark sky, remember that the typical 'one-hour' processor if you are using print film or the typical laboratory if you are putting transparency film in for processing usually deal with everyday subjects where there is no doubt about the edges of individual frames. This is not the case with eclipse transparency film where the Sun will be surrounded by total blackness during the partial phases as a result of using a filter—and surrounded by totally clear film in the case of the processed negative film. It is advisable in the case of both types of film to shoot a first picture of an ordinary subject on a new film so that the automatic printing machines in the case of negative film and the automatic cutting and mounting machines in the case of transparency film can be set up with the Sun centrally positioned in each frame. In fact, where transparencies are concerned, it would be safest to have the film returned processed but uncut and unmounted—and for you to cut up the transparencies yourself and mount them. You can only blame yourself then if it is done wrongly!

Filters

It is important to emphasize that whether observing only or taking photographs, safe protection must be used at all times other than during the period of totality. The two main types of filter were described in Chapter 3. Mylar will be readily available from distributors. It is easy to use and can be purchased in a number of sizes as well as ready mounted in cardboard holders to fit the proprietary filter holder systems like Cokin. Prices are not too extreme. A 100 mm square of unmounted Mylar costs a little over £25 and mounted forms to fit filter holders are about £35 or £40. The drawback is the light blue colour of the image both as seen with the eye and recorded on film. The former does not matter very much but I do not like transparencies or prints showing a blue Sun! This can be corrected reasonably well by using a Wratten type 21 (orange) filter which adds just a little more complexity as well as cost—a 75 mm square filter costs between about £11 and £25 according to brand—but is well worth considering. The glass filters coated with metallic alloys are far more durable than Mylar, are of at least as good optical quality if not better, and produce a yellow Sun. They are inevitably more expensive but the difference is not enormous. A mounted glass solar filter of up to 90 mm in diameter is of the order of £65 to buy. (Of course, you might be able to borrow a welder's glass from somewhere. I have never used one but I strongly doubt whether the optical qualities will be good enough for serious photography.)

It must be for the individual to decide but I confess that I almost always use glass, alloy-coated filters for all of my 'white light' solar work. The important thing is for you to decide quickly which filter you want to use. There are enormous runs on filters and other aids when an eclipse approaches and it could well be that if left to the last moment the type that you decide to use is out of stock.

A third type of filter was briefly mentioned in Chapter 3—photographic neutral density filters of lacquered gelatine, which also yield a yellow solar image. I have used these from time to time but extreme care needs to be taken because, although they absorb visible radiation well, they transmit infrared and ultraviolet. Using these is a two-stage operation: a lens is focused and the Sun framed using, say, the two-thickness negative sandwich already described and then that is removed, being replaced by the neutral density filters. I say 'filters' advisedly because the densest of the filters has a value of 4.0 (which represents a filter factor of 10 000). Solar photography demands an ND value of 5.0 (factor of 100 000) or of ND 6.0 (a factor of 1000 000) so it is necessary to add an ND4.0 filter to an ND1.0 to get the former value (or a 3.0 to a 2.0) and an ND4.0 to an ND2.0 to get the latter. While cheaper brands are available, I have always used Kodak Wratten ND filters and while excellent results have been obtained—the time lapse images which appear in figure 7(*a*) were created using Wratten ND filters—the prices are not low. Given the need to purchase two appropriate filters as well as to mount them, the end cost is likely to be almost as high as that of Mylar or glass solar filters, so either of the latter two would seem to be the most sensible choice. One final note of caution: remember that the filter always goes on the front element of the lens where it must be very securely mounted or taped. If at all possible, I strongly recommend that you have your filter custom mounted so that it fits securely on the front of the lens.

(*a*) The Telephoto 'Patrol'

So the equipment is all prepared. You have an SLR with a telephoto lens mounted on top of a tripod. The lens is filtered and a shutter release cable is also available. First of all, it is essential that you practise what you hope to do on 11 August. I am assuming that you intend to take a series of telephoto shots during the partial phases of the eclipse as well as during totality.

If you are in Cornwall or Devon on eclipse day, the eclipse will start with the Sun a little south of due east at an altitude of around 36°. It moves about 15° east–west in one hour and by the time of totality it will have climbed to about 46° altitude with a further climb to 53° or 54° at the end of the eclipse. (If you find this difficult to envisage, hold your arm out fully in front of you and clench your fist. Rotate the knuckles until they are vertical and the distance between the lowest knuckle and the highest is roughly 10°. So, if you imagine five of your fists on top of one another you will have a very rough idea of the height of the Sun at totality.) The eclipse will last for just over two and a half hours in all with totality lasting for a little over two minutes on the centre line.

At least fourteen days before 11 August (earlier if that is impossible) you should conduct a serious practice. If weather allows, run the practice at the same time of day as the eclipse—from approximately 10.00 am BST to 12.30. With filter safely in position start making exposures with the Sun in the middle of the frame. I would suggest that you use ISO100 speed film and that with the lens set at f/11 you expose three frames at 1/125, 1/250 and 1/500 of a second respectively at 15 minute intervals. After the practice, you will be able to see from these exposures what will be the optimum exposure for your set up on 11 August. You do not need to be too precise with the timing so a watch would do fine or an alarm timer of some kind if you prefer it. Incidentally, the partial phase exposures will not alter during the practice nor on 11 August as the eclipse progresses: the exposure for the Sun when the first small chip is out of it is exactly the same as when only a thin sliver of the photosphere remains just before totality.

Quite apart from valuable exposure guidance, this practice will show how easily or otherwise you can track the Sun. The longer focal length you choose, the faster the Sun will appear to move in the viewfinder. You will need to monitor its position continually and to make adjustments to the lateral (azimuth) and vertical (altitude) controls of the tripod head at frequent intervals. Can you make these alterations easily without 'losing' the Sun? Can you lock everything firmly after you have made an alteration?—because the angle of the camera/lens combination will become steeper as the time progresses. Does the weight of camera/lens keep taking the Sun towards one side of the frame as soon as you release the controls? If it does, you'll have to learn how to position the combination so that the Sun appears at the top edge of the frame and then drops to the centre after you release the controls. Better to do all of this well before rather than to encounter the problems on the day. Of course, you will not be able to practise exposures for totality, but at some time before or after your test run and with the camera pointing away from the Sun, see how well you can remove the filter without disturbing the camera on the tripod. It should come off as smoothly as possible but not so easily that there is any danger of it falling off during photography of the partial phases. Two further points. For your practice and particularly for your adventure on 11 August when you will be in a relatively strange location, make sure the tripod is set on very firm ground (not sinking in soft turf with the weight) and that the tripod head is as level as possible; you'll be altering it in two axes but the third should be fixed. And, if your practice day and/or 11 August are hot, cover your camera and lens with a light cloth to keep them as cool as possible.

With lessons learned you will be ready for the day itself. So far as your programme is concerned, let me emphasize again the point already made about taking plenty of time to look at what is happening during the eclipse and around you. The hour or so before totality is a relatively leisurely time. One or two exposures roughly every fifteen minutes, which you should bracket, are about right although if there are interesting and large sunspots on the disc you might want to take photographs with the Moon's leading limb or edge just approaching them and then immediately after they have disappeared. I always make a note of the time that exposures are made—a data-back which prints information on the film is very useful for this although you may wish to take notes or use a small tape recorder.

As totality approaches things will become more tense. If you have been exposing many frames and are nearing the end of a film, well in advance of totality—I suggest a minimum of ten minutes—put in a new film and get firmly set up again. Also, as totality approaches I would advise you to establish a very firm exclusion zone around your tripod and camera because you do not want somebody to knock the camera off its alignment on the Sun nor to kick the tripod over. Close to totality this could be a disaster and even outside of totality you would be surprised at how difficult it is to find the Sun with a long lens when a solar filter is in position and your tracking has been disturbed. (I never mind showing people the partial phases in the viewfinder but during totality my area is forbidden territory!)

Look back in Chapter 2 at what happens during totality and plan a limited programme of exposures. As soon as the filter is removed, as the last of the photosphere disappears (do not be too quick!), I would recommend you take no more than three exposures to cover the hoped-for appearance of Baily's Beads, Diamond Ring and prominences at the Moon's leading limb. Then, enjoy the sight for thirty seconds or so before shooting possibly five frames at different exposures to record the corona. Pause to look at things once more before shooting two or three frames at faster shutter speeds again to get the edge phenomena at the Moon's trailing limb this time, after which the filter has to go back on as the photosphere reappears. One important thing is not to rush the exposures: let shutter vibrations die down before exposing another frame otherwise you risk image smear. Guidance on the exposures to try are given in table 5.1: you can make your own selection.

Table 5.1. Solar eclipse exposure guide.

Film speed (ISO)	Partial phases (ND5.0 filter)	Totality				
		Baily's Beads	Diamond Ring	Prominences	Inner corona	Outer corona
25	1/60th at f/8	1/250th at f/ 8	1/60th at f/8	1/125th at f/4	1/30th at f/2.8	1/4s at f/2.8
50/64	1/125th at f/8	1/500th at f/8	1/125th at f/8	1/60th at f/8	1/30th at f/4	1/4s at f/4
100/125	1/250th at f/8	1/1000th at f/8	1/250th at f/8	1/125th at f/8	1/30th at f/5.6	1/4s at f/5.6
200	1/250th at f/11	1/1000th at f/11	1/250th at f/11	1/125th at f/11	1/30th at f/8	1/4 at f/8
400	1/500th at f/11	1/2000th at f/11	1/500th at f/11	1/250th at f/11	1/60th at f/8	1/8th at f/8
800/1000	1/500th at f/16	1/2000th at f/16	1/1000th at f/11	1/500th at f/8	1/125th at f/8	1/15th at f/8

NB: Use suggested partial phase exposures as a basis for early tests. Bracket exposures for totality phenomena by at least one shutter speed/stop either side of suggested values.

You will find that the two minutes of totality will go in a flash. You might want to have a tape recorder running quietly with a pre-recorded tape alerting you to the passage of time or you might have a sophisticated watch that can sound an alarm at intervals. Light levels fall off quite sharply during totality so you might choose to employ a small, shaded torch or (like an experienced astronomical observer) a red light torch to enable you to see the shutter speed settings when altering them. If you want to provide yourself with a checklist of different shutter speeds for your totality programme, write the notes down in large capitals with a black felt pen on a card. Try to avoid altering the aperture settings (as distinct from shutter speeds) as much as possible during totality: you'll have quite enough to do in any case and will risk altering the focus of the lens. Afterwards you will find that you cannot remember quite what you did or did not do during totality, which is where a data-back can prove valuable. Then you can go on photographing the partial phases of the Sun as the Moon begins to move off the disc—or relax and enjoy things.

The above patrol idea can be pursued, of course, by anybody with the appropriate equipment outside the zone of totality. The only difference is that the filter remains in place all of the time and that, instead of images of totality, one or two frames are taken of the solar disc when the partial phase is at its most extreme—over 99% obscuration in the case of my home area in Havant, Hampshire.

(*b*) Time lapse image

There is an alternative project to the series of telephoto images of the Sun and this too requires careful planning: it is to image the entire eclipse on one frame of colour film. Fortunately what happens during the eclipse allies itself to the use of standard photographic equipment with nothing out of the ordinary involved. The whole of the eclipse—partial phases and totality—will last just over $2\frac{1}{2}$ hours. During this time the Sun—travelling at the 15° per hour noted above—will have moved therefore by about 37.5°. The 50 mm focal length standard lens on many 35 mm cameras has an area of 27° in the vertical by 40° in the horizontal but the diagonal of the lens is 46°. So although the Sun climbs at the time of morning the eclipse will occur as well as moving laterally by the amount indicated, it so happens that the diagonal of the 50 mm focal length lens is adequate to record the entire eclipse with something in reserve—in fact approximately three hours of solar movement will just fit within the film frame. It was shown in the previous section that a single 50 mm lens image of the Sun is very small but a sequence of them in a 35 mm film frame which can be projected as a transparency onto a screen or enlarged as a colour print from a negative is impressive.

A test run taken in August 1998, almost exactly one year in advance of the eclipse is reproduced here (figure 7(*a*)). Throughout, the 50 mm lens was stopped down to between f/8 and f/11 and a shutter speed of 1/250th of a second was selected using Ektachrome Elite II ISO100 film. Neutral density filters were used in this case and an exposure was made every 10 minutes using the interval timer on my Nikon F4S. The multiple exposure lever, however, had to be operated manually before each of the programmed exposures. A second image (inset) shows the effect of an exposure every 5 minutes during a shorter test run.

Once again a thorough practice a few weeks before the eclipse at about the same time in the morning will prove very valuable. Although you will not be putting a heavy telephoto lens on your tripod it is absolutely essential with this time lapse image that the camera does not move by the slightest during a long series of exposures. And of course you must be able to override the film advance of your camera in some way. The Nikon used in my test has an easily moved lever for this purpose but if this facility is not available the same effect can be achieved by depressing the film rewind button normally located in the base plate of most 35 mm cameras—another reason for making sure the camera is very securely fixed on the tripod head and that you have access to the button.

The method of proceeding with the practice run initially and then, all being well, with the sequence on eclipse day is very straightforward. With the camera loaded with your chosen film (you can easily work out the exposure variations if you choose a film of different speed from the one I used), the solar filter securely in place on the 50 mm lens which is focused and taped on infinity, and a cable release fitted, thread the tripod head bolt into the base of the camera. Adjust the tripod, with its legs securely tightened, so that the camera faces in the direction of the Sun. Level the tripod head but with both the azimuth and altitude bolts of the tripod mount slackened off, adjust the camera direction and angle so that the Sun appears as close as possible to the lower left corner of the frame (lower right if you are using a right-angle viewfinder). Lock both the tripod head bolts and you should be ready to proceed. Whether you make exposures every five minutes or ten you will need some form of timer to accurately measure the intervals. You should also adopt a routine from the beginning that you follow without fail. After each exposure, immediately depress the film rewind button if this is how you are overriding the wind on mechanism and tension the shutter (assuming you are not using a motor drive)—yet one more reason for making absolutely sure that the camera is firmly positioned on the tripod head. It is very easy to get confused about whether you have depressed the rewind button or tensioned the shutter but if you do it immediately after each exposure as a routine there should be no problem. As in the case of the telephoto programme, if it is a hot day on 11 August 1999 place a cover over the camera between exposures to keep it cool.

There is only one other small complication. Although this is of no consequence in your practice run, on eclipse day you will obviously want the middle image of the sequence to occur during totality. Therefore you must count back from that time in ten or five minute intervals to establish when you should begin the sequence. Appendix A gives information for locations on the track of totality in Cornwall and Devon: let us take Redruth as an example. Mid-totality is at 11.12 BST there so if you choose an evenly balanced sequence of, say, seventeen ten-minute interval exposures (which gives you small reserve of space on the film frame) you would expose the first one at 9.52 BST before the eclipse has begun. The times would run thus:

09.52, 10.02, 10.12, 10.22, 10.32, 10.42, 10.52, 11.02–11.12 (totality)–11.22, 11.32, 11.42, 11.52, 12.02, 12.12, 12.22 and 12.32 (just as the eclipse ends).

Of course, for the one exposure at mid-totality—for which you can derive exposure advice from table 4.1, making the change to shutter speed only for safety's sake—you need to remove the solar filter. Replace it as soon as you have made the exposure, and prepare for the next one using the procedure already recommended.

The time-lapse image is also one that can be repeated outside of the totality track, the middle image in this case being the maximum extent of the eclipse from wherever the sequence is being shot. Frankly, I think that here the timing sequence is somewhat less critical because a few minutes either way is not going to make an enormous difference to the extent of the solar crescent left around mid-eclipse unless one is wanting to be very accurate. Timings for various locations outside the track of totality are given in appendix B.

Some further points about the time lapse project. The images of the Sun will be bright against a totally black background. A successful outcome will be an achievement for you but you may wish to add further interest by having a backdrop for the sequence composed of a scene at or near the location where you are taking the image. (Select an attractive one!) You could take a (somewhat underexposed) picture of the scene on the same frame of film either before or after the sequence is obtained but I think this is risky because of potential exposure problems and also the need for very precise positioning of the foreground. It will be far better to take one or two landscape scenes quite separately and then produce a composite by purely photographic or digital means afterwards. Remember that your sequence shot with the 50 mm lens will be filling the film frame from one corner up to the opposite corner so to fit the sequence into the sky of a landscape background and allow for some foreground interest you will need to adjust the scales between the two frames. Increasing the area of the landscape scene and dropping the sequence into it would be the easiest way.

Secondly, there is no need to stick to the 50 mm lens as your choice provided you realize that, say, a 200 mm focal length lens producing an image of 2 mm in diameter will not allow you to cover the entire eclipse with time intervals in between each exposure. Figure 7(*b*) was taken as a test toward the end of August 1998 using a 180 mm lens at 5 minute intervals. Notice that, after leaving a small margin at the commencement of the multiple exposures, there was room for ten exposures. An even number would not allow the image taken at totality to be balanced by those on either side so using this focal length of lens or one close to it nine exposures should be made, with a slightly larger margin being left at the edge of the frame before the sequence is begun. Again, using Redruth as the example, on 11 August 1999 the timings for a sequence of nine exposures would be:

10.52 BST, 10.57, 11.02, 11.07–11.12–11.17, 11.22, 11.27 and 11.32.

The eclipse at this location begins at 09.57 BST and ends at just gone 12.32 so you might consider this would truncate the partial phase coverage too much. If so, an intermediate position with a lens of around 100 mm focal length could easily be worked out.

Finally, I am not a pessimist but at every solar eclipse I always believe in having a camera in reserve during totality in case I have a major problem with my prime equipment. I have done so ever since the eclipse in the Philippines on 18 March 1988 when a main camera problem was compounded by weather conditions which resulted in the Sun only appearing from behind clouds in brief snatches. I had a second Nikon with the 180 mm f/2.8 lens mounted on it and this enabled me to get some quite reasonable images even though they were inferior to those few taken with the 1000 mm f/11 Nikkor. One of them is reproduced here (figure 8) You may wish to follow the same practice.

(*c*) 'Atmosphere'

The previous chapter elaborated the considerable number of images which can be taken about the theme of the eclipse even though, except in the case of the crescent-shaped images being projected during the partial phases, none was directly astronomical in character. However, there are a number of images which can be taken that are—besides the projected crescents, there is the lunar shadow advancing from the west (figure 9) as well as the elusive phenomenon known as shadow bands which can appear as alternating light and dark bands of light rippling across any light surface in the seconds before and just after totality. Pictorial images that capture the mood of an eclipse is another theme (figures 10(*a*) and 10(*b*)). Then there is the period of totality itself which could easily be photographed with a hand-held camera either with a wide angle lens taking in much of the scene or a moderate telephoto recording the immediate area of the eclipsed Sun together with Venus and Mercury.

I am very aware that keen photographers (hopefully) witnessing a solar eclipse for the first time should not attempt to do too much: time must be left simply to look at and relish the magnificent spectacle in the sky. If you decide to attempt the programme of telephoto images of the partial phases and totality (although there is nothing to stop you concentrating entirely on totality) my advice would be to attempt nothing else. That programme on its own will be demanding enough. In any case, a companion who is also interested in photography could take care of all the secondary themes. However, while the time lapse exercise demands attention to detail and care, it is not so intensive as the telephoto programme and, with just one image due to be taken at totality, the use of a second camera to record the lunar shadow advancing and then hand-held shots of the Sun in eclipse would not be too demanding. Naturally, there is another possibility: that you ignore the projects outlined at length here and simply enjoy yourself with or without a hand-held camera during totality!

The idea of hand-held photography during totality, but using superior equipment to that used by the casual snapshotter, would appeal to me if I were not having to photograph the eclipse professionally. Despite the view expressed earlier about film, I would load a good 35 mm SLR with one of the recently introduced colour negative films which can be rated for individual exposures at anything from about ISO100 to 1000 and yet which has excellent quality. I would select a zoom lens but not one of extreme focal length—possibly one offering a 24–120 mm focal length range or 35–135 mm. I would keep the focal lengths relatively short for two reasons. Firstly, in my view the old-fashioned advice that when hand holding a camera, you should never use a shutter speed slower than the reciprocal of the focal length of the lens being used is very sound. Thus, with the upper limit of the two lenses I have selected above, one could with reasonable care get away with shutter speeds of 1/125th. Secondly, as we have already seen, the Sun will be quite high at totality at about 46° and therefore to be able to include it with a margin of sky above and also to include some foreground, a wide angle lens will be needed—a 35 mm wide angle lens, for example, offering a vertical of 55° with the camera used upright, and a 24 mm focal length lens considerably more. A problem with the latter will be the even greater reduction in the size of the Sun, but in this concept the subject is the Sun in the sky with planets and stars—not just the Sun. Any exposure which shows the corona will enlarge the apparent size of the Sun in any case.

The attraction of the zoom is that with just one lens one could take the whole scene with a foreground to give the sky context and then zoom in to concentrate on the sky and the objects in it (figure 2). The distance between Venus in the east (which will be extremely bright and unmistakable) and Mercury on the west, with the Sun almost centred between them, is about 33° so this would be covered using a focal length of around 50 mm. Mercury will not be as bright as Venus and the next brightest object after Venus will be Sirius below and to the south west of the Sun. Increasing the focal length to around 90 mm would enable Venus and the Sun to be covered and then the full focal length would allow concentration on the eclipsed Sun.

As to exposures, if one of the versatile negative colour films is selected, I would use the bottom line of table 4.1 as a basis. The zoom lens is likely to have a maximum aperture of between f/3.5 and f/5.6 according to focal length chosen but I would keep the aperture at f/8. This should give some coronal detail even at 1/125th second with the other phenomena being exposed at higher shutter speeds. Of course, if you decide to take this path but put the camera on a tripod then the quality of all of your shots would improve and you will have a far greater chance of getting a better exposed foreground with the wide angle—and a much more extensive corona around the Sun when you zoom in—because you will be able to choose slower shutter speeds with confidence.

In fact, one might almost say the sky will be the limit! Good luck.

Chapter 6

The Moving Picture

Camcorders are very popular today. As in still photography, there are innumerable owners who have the camcorder simply to record holidays and other family occasions, as well as a much smaller number for whom it is the means to a fascinating and serious hobby. Both groups will doubtless be well represented in Cornwall and Devon on 11 August 1999 and guidance on an event that very few (if any) will have seen or attempted to record before is therefore appropriate.

Camcorders have developed greatly over the past decade. New models with increasingly sophisticated features follow each other on to the market within months, if not weeks. There are a number of different formats including 8 mm, Hi8, VHS (several types) and the more recently introduced DV (digital video). To attempt to cover all of these is impossible and in any case is not required. Basic photographic principles apply to much of what the camcorder will be faced with in August 1999. Further, I must be honest and admit that I am not normally a camcorder user. However, I have applied my photographic knowledge and eclipse experience to using a camcorder for a sustained series of tests and can therefore give valid guidance. The camera I used was not an up-to-the-minute model; indeed it could be described as 'elderly', being a Sony CCD-F550E. This is an 8 mm unit with an 8× zoom and a manual override for both focusing and setting the shutter speed. No modification to the iris is possible. The manual control capability is important but other features of the camcorder will have been superseded by many of the later models, of which there are more than 70 currently on the market. Nonetheless, it was a useful and reliable test vehicle.

It is essential that the camcorder user read the earlier chapters of this book to establish what happens during the stages of an eclipse and also to absorb the guidance given to still photographers, much of which will apply to recording the moving image, too. As in those chapters, I identify three themes.

Pre/post eclipse atmosphere. These events present no unusual problems for the camcorder user. Videoing the various scenes is conducted in normal daylight and obtaining a pleasing record is down to the talent of the camcorder user, who can be operating the camera in the automatic mode at this time.

Partial phases. Any camcording of these must be conducted with the appropriate solar filter in place over the lens. Since the Sun moves, as already indicated, about 30 times through its own diameter in an hour, it is essential that the camera be mounted on a firm tripod for tracking. It would greatly facilitate the smoothness of this operation if the tripod were to be equipped with a suitable pan/tilt head.

The *size* of the solar disc that will be obtained is obviously dependent on a particular camcorder's zoom capability. The now somewhat 8× limited zoom on the unit I used (without any attempt to boost the focal length) yielded a solar disc of about 2 in (5 cm) diameter on a 26 in television screen. This was fairly small but entirely adequate to secure information about focusing and exposure accuracy. Current camcorders usually have optical zoom factors of at least 16× and often 21 or 22×, so these will yield very satisfactory results as regards disc size. All of this can (and should) be checked well in advance of eclipse day.

If manual control is available, *focusing* presents no problems—the lens is simply set on infinity and left there. A recent review of all the camcorders currently on sale revealed that fortunately all but about ten (out of more than 70) could be operated on manual focus. This is important because the F550E when set on auto focus frequently 'hunted' for the filtered Sun and when it did come to rest it was considerably short of the infinity setting. Experience of still automatic cameras is that they frequently yield better results than theory predicts they should. That may be the case with auto-camcorders but the pre-eclipse tests will reveal whether there is any point in pursuing matters further with an auto-mode focusing camcorder if good results are sought.

Nor is focus the only problem. Exposures of the Sun through a Mylar filter is another difficulty facing auto-camcorders, particularly as the already quoted market survey reported that around one-half of the current camcorders allow no manual override of exposure controls. Images viewed on the TV screen after my tests of the Sony in auto-mode appeared at first sight to be reasonably satisfactory. But when they were compared with images obtained using manual exposure settings, they were found to be about three-quarters of an inch (2 cm) greater in diameter. I concluded that this was the result of a combination of overexposure and poor focusing. This might be acceptable to some users but on this evidence no sunspots, which might well exist on the disc at the time of the eclipse next year, are likely to be seen clearly when reliance is placed on automatic mode settings.

The Mylar filter when used with the camera in manual mode yielded not a blue but a whitish (slightly over-exposed) disc at the Sony's fastest shutter speed of 1/4000th second. While additional expenditure is involved, I would strongly recommend the addition of a Kodak Wratten 21 filter (or its equivalent from another manufacturer) to the Mylar for two reasons. It yields a much more normal yellow/orange image of the Sun, absorbing strongly as it does in the blue and blue/green areas of the spectrum. As important is the fact that, because of that absorption, the camcorder sensor does not receive as much light *per se* so that the overall exposure level of the image is improved. (For those who might want to be more artistic, shutter speeds of between 1/250th and 1/1000th in my tests yielded a pleasing sky background behind an overexposed Sun —blue in the case of the Mylar filter used on its own and a splendid orange/red when the 21 filter was added.)

The individual can verify these various points in advance using the equipment that is being proposed for the eclipse. The question to be answered in contemplating the eclipse is what exactly to do. Although battery life is an important consideration—and a fully charged spare should be available—it is possible to contemplate tracking the Sun smoothly during the entire period up to and including totality and beyond to the end, in which case several tapes would be required. Here it would be best if the camcorder could be 'piggy-backed' on a telescope or camera mount tracking at the solar rate! However, this would make for a rather boring record since the encroaching Moon moves almost imperceptibly over the solar disc and I cannot imagine anyone sitting down afterwards and watching a television screen for the $2\frac{1}{2}$ hours or so of the entire eclipse. I may be wrong.

Much more interesting will be the creation of a time lapse record. Most of the contemporary camcorders have a facility for taking still images. We can imagine a number of frames being exposed every thirty seconds or one minute and then an entire sequence being replayed at a suitable speed. The important thing, however, will be for the Sun to be centred very precisely in the still frames because if it is not the time-lapse record will be continually 'jittering'.

The recommended pre-eclipse tests will yield another very useful piece of information about the totality period of the eclipse. My tests with 8× zoom demonstrated that the Sun moved entirely across the view in the finder from lower left towards upper right (no reversing of the image with camcorders!) in somewhat over 15 minutes. Since totality lasts for only a few seconds more than two minutes at best, this means that even with the more extensive zooms now available a Sun that has been reasonably well centred in the viewfinder just before totality begins will incur no risk of running out of the frame of view of a static, tripod-mounted camcorder. In choosing a zoom factor, bear in mind that, when the eclipse actually takes place, the solar corona during totality can be extensive and that, while it can vary somewhat from eclipse to eclipse, it can spread across two solar radii or more on each side of the Sun. This should be allowed for in selecting the zoom factor since it would be a great pity to have prepared well in advance and then to select a degree of zoom which has to be hastily corrected during totality with the inevitable and distracting movement of the camcorder.

So totality arrives. Here, the camcorder will have one great advantage over the still camera—sound recorded at the same time as the visual record is being made, and high quality sound at that. Without the need for any additional equipment, the camcorder will capture the gasps of amazement, the awestruck comments and the shrieks of delight from the assembled throng (besides your comments) as the corona comes into view. That will add greatly to the impact if the attempt to tell a story or to portray the atmosphere during totality is being made.

But the visual record is of course the prime task. It is usually stated that in preparing to record totality during a solar eclipse (with the filter used during the partial phases removed, of course), prior exposure tests should be conducted using a full or nearly full Moon as a subject. Mine using the F550E in manual mode on a Moon two days past full recorded the best exposed lunar disc using 1/4000th second once again—although without a filter in position this time. Some detail on the Moon was seen with the shutter speed set at 1/2000th second but maximum detail of both the lunar maria and highlands was undoubtedly recorded at the faster shutter speed.

Tests were also conducted to record general scenery and the sky during mid-to-late twilight on a late summer's day, at between 20.00 and 20.30 BST. Good sky as well as foreground detail could be seen at the camcorder's slowest shutter speed of 1/50th second set manually. That same shutter speed, with the zoom lens altered to the maximum 8×, secured images of two of the brightest stars of the summer sky—Vega in the constellation Lyra (magnitude +0.03) and Arcturus in Bootes (magnitude −0.04) quite well.

The results of tests like these should help the camcorder user to plan for eclipse day. But I foresee a major problem resulting from the briefness of totality. With a reasonably long totality, the following sequence of coverage could be envisaged by somebody who was determined to maximize the returns.

(i) With the lens set to wide-angle and filter removed, recording the immediate pre-totality atmosphere such as the Moon's shadow rushing in from the west.

(ii) The controls of the camcorder are then altered to high shutter speeds with iris stopped down and highest zoom factor set, preparatory to recording the immediate pre-totality phenomena such as Baily's Beads and the Diamond Ring. Just before these occur, the camera is pointed at the Sun and locked into position.

(iii) These phenomena are recorded for several seconds. The risk of over-exposure at this time could be countered by the use of a relative low density ND filter on the lens, perhaps an 0.5 or a 1.0.

(iv) If used, the ND filter is removed and exposure controls altered to record the corona. Exposures are varied in an endeavour to secure inner and then outer corona.

(v) Zoom factor is reduced and exposure increased to record stars and planets surrounding the eclipsed Sun. (Based on my tests, the very bright planet Venus and nearby stars such as Sirius, Procyon, Rigel and Betelgeuse should be recorded. The much dimmer planet Mercury and stars like Castor, Pollux and Regulus are a little more problematical. See figure 2.)

(vi) The zoom factor is again increased and the ND filter re-installed for imaging Baily's Beads and Diamond Ring at the end of totality.
(vii) Afterwards, the ND filter is removed and replaced by the Mylar filter.
(viii) Apart from the above, enjoy just looking at the spectacle.

With a totality of just over two minutes, this will be totally impossible on 11 August 1999 even for a highly experienced, solar eclipse camcorder specialist. However, aspects of this programme could be carried out by several camcorder users making themselves responsible for different sequences—and certainly by the members of a video club endeavouring to make a film about the eclipse.

For an individual determined to cover some sequences of the eclipse but to leave some time for just enjoying the sight, the following programme could be a useful basis.

(1) The camcorder is mounted on a tripod and the Sun tracked during the partial eclipse phases. With zoom and (manual) exposure controls set at factors established by earlier tests, record the solar image for perhaps 30 seconds every five or ten minutes. Use a Wratten 21 filter for improved results.
(2) Just before totality and with the camcorder still running and Sun centred in the viewfinder, stand in front of the lens and carefully remove the solar filter. Continue to stand in front until Baily's Beads are seen. Step aside. Check that the Sun is still in the viewfinder!
(3) If your camcorder allows you to alter exposure controls by remote control, try short sequences either side of what you considered to be the optimum setting as established in pre-eclipse tests if this latter setting appears to be under or overexposing as seen in the view finder. If the setting appears to be accurate, leave well alone.
(4) Let the camcorder record the last seconds of totality and then stand in front of the lens once again. Restore the solar filter and continue with coverage of the partial phase sequences if you decide you want to do so.
(5) There is of course no reason why you should not ignore the partial phases and concentrate solely on totality. The camcorder, set as in (1) above with Sun centred but minus the filter, could be started just as totality begins and then allowed to run whilst you just enjoy the sight with your eyes.

Beware of one thing if you are using a Mylar or other filter during the partial phases. I have seen it recommended that the problem of removing the filter whilst the camcorder is running (with the risk of knocking it out of alignment or at least causing a case of the 'jitters') can be countered by making some form of filter holder on a stand which is in front of the lens but not touching the camcorder. This will not work. If there is any gap between the front of the lens and the filter, so that light is admitted, reflections off the filter are virtually certain to ruin the picture.

Over the years, I have seen a number of videos made by solar eclipse expeditions that made extensive use of camcorders. I have found these to be generally weak in the following respects.

- The camcorders are all too often hand held with images moving and sometimes violently so. Even the units mounted on tripods are subjected to considerable movement because of alterations to zoom and exposure controls at the camera. (I would assume that the increasing sophistication of remote control and the new camcorder image stabilization systems will help considerably here.)
- Altering the degree of zoom during totality is disconcerting. It seems to be done usually to make the solar disc appear as big as possible. Having fixed on an optimum size to allow for the spread of the corona, the zoom should not be altered.
- CCDs and other digital sensors have a very great dynamic range but there invariably seems to be a degree of overexposure during totality sequences. I cannot remember a single video sequence in which I have seen a well saturated red prominence or section of chromosphere during early or late totality. If totality were to be long enough, a careful selection of ND filters

would probably solve this problem—but, where this is not possible, might not post-eclipse editing and digital enhancement be able to render these phenomena closer to the impactful detail seen in so many still images?

As with still photography, the demands of photography during the eclipse —and particularly with totality being relatively short—may cause the camcorder user simply to look and enjoy. But one can have a go for part of the eclipse and look for the remainder. And if you follow that course and also have a purely automatic camcorder, use that too subject only to the demands of eye safety outside of totality. While manual control undoubtedly yields the best results, I am pleasantly surprised by what camera manufacturers and the makers of sensors and film can achieve with fully automatic systems at least some of the time.

Chapter 7

The Digital Path

Computers in the home are now commonplace and increasingly programme packages include one for changing or enhancing scanned images. Digital cameras as such have not yet been taken up by the snapshotter but manipulation of images at home or as offered by a growing number of commercial outlets is likely to mean that the eclipse of August 1999 may well provide the occasion for subsequently rescuing or improving pictures taken on film. Hence the addition of this brief chapter to a 'primer' on eclipse photography. Three examples of digital treatment from the simple to the more complex follow.

The use of Mylar has been frequently mentioned in earlier chapters. The blue image that the filter yields can be corrected in camera by the use of an orange filter but it is easy to correct it digitally, too. The original blue image reproduced here was taken in July 1990 on a high resolution ISO100 colour transparency film when the Sun was still showing great activity as can be deduced by the number of sunspots visible. Much more recently this image was scanned using an Epson FilmScan 200 unit and manipulated in Adobe Photoshop. The scanned image was loaded into the computer and, because in altering the colour content of a picture of this kind it is best to avoid tinting the background dark sky when working on the main subject, Photoshop's magic wand tool was used first of all to select the entire sky background and then, by clicking the inverse command, to select the solar disc. This means that any subsequent manipulation would only affect the disc. The Hue and Saturation command within the Image/Adjust menu of Photoshop was then chosen and, with the Colorize and Preview boxes of the command panel ticked, the hue of the selection was set to 31, the saturation to 69 and the lightness to minus 15. The effects of various settings can, of course, be seen on the PC monitor and a satisfactory result saved and printed out.

The resulting image is shown here (figure 17) beneath the original. The colour balance is quite good although some contrast has been lost in the sunspots. Photoshop is a complex and enormously versatile package and I am quite sure that further improvements to the modified image could be achieved by further work, perhaps using other tools. A third image is that of one of the images taken at the same time but through a metal alloy coated glass filter. The sunspots are as clearly reproduced here as in the Mylar image but too much should not be read into the different yellow colour of this image compared with the digital conversion: glass filters coated with metal alloys can vary greatly and I have another filter of this type which produces on film a colour balance much like that in the digital image.

On 24 October 1995 a short total solar eclipse occurred over much of Asia and the Pacific. The weather prospects were good in north west India in the Great Indian Desert not far from the border with Pakistan and that is where I covered it. Inevitably, during the trip to India a visit was planned for our party to the magnificent Taj Mahal outside of Agra and, while totality would not occur in the sky above it (the eclipse was about 99.7% there), before I went I envisaged the subsequent creation of a composite image in which pictures of the eclipsing Sun would be placed in the sky above the mausoleum. Of course, in placing the resulting composite in the photo library the caption stated that totality was not seen from the Taj Mahal and that it was an attempt to combine one of the most beautiful works of humankind with one of the most magnificent sights of Nature. It would certainly have been possible to create the composite by purely photographic means but the result can be accomplished much more easily and therefore quickly using digital techniques.

The composite which is reproduced here (figure 18) can best be regarded as two separate pictures. Let us take the Taj Mahal first. I have mentioned elsewhere the fascinating light which is seen close to totality—a grey-blue to my eyes usually—and to simulate this I deliberately took to India several roles of transparency film balanced for use in tungsten lighting. This film is intended to maximize the blue region of the spectrum (which is relatively weak in artificial lighting) and when used in daylight results in the pictures having a marked blue bias. While such a balance could be achieved with digital modification I think it is always sensible to create in an initial film image any effect which can be achieved in camera without any fuss. The result was a quite reasonable approximation of the lighting at totality and no additional work was required in this direction.

The chosen slide was then scanned and the data loaded into the computer using Photoshop. The picture was cropped to achieve a pleasant composition in which the Taj Mahal was balanced by two turrets on either side seen across the tops of trees. Several blemishes on the scan and also one or two intrusive objects on the pathways (such as paper dropped by visitors) were removed by using the rubber stamp tool—an extremely valuable tool which takes a sample of an image or part of an image and enables you to apply that sample to a different image or part of the same image. Thus in this case, by clicking on a clear area of the path I could 'paint' it over an adjacent object.

The picture generally was satisfactory but when it was exposed the Sun was behind me and to one side. This created heavy shadows in the galleries at the front of the building. These needed to be lightened and again this was straightforward in Photoshop though a little patience and care are needed. Each area of shadow was selected using the pen tool which permits you to draw lines with great precision. When the outline of a shape is closed (like a fenced-off field) the path thus created can be made into a selection—the selection then being modified using any of a multitude of tools and techniques. To lighten the shadows I used the dodge tool (like the technique of holding an area back in photographic printing) which could be moved over the selected area without any effect on the surrounding area. Each of ten hard shadow areas was lightened in this way, as well as the smaller number of shadows being reflected in the water. The modified image was then saved.

It was then time to turn to the solar images. I thought that an in-camera time-lapse image of the entire eclipse would not be too impactful in the composite because the solar discs would be relatively so small and there would be too many of them. I therefore decided to compose a sequence of eclipse images from the numerous individual pictures I was shooting using a 1000 mm f/11 Nikkor lens. This enabled me to limit the number of images and also to choose discs that were taken at approximately the same time before and after totality. Of course, one image of totality was included. All needed to be carefully aligned and spaced on a black background and to do this a grid that can be called up as a temporary overlay was very useful. In compiling the discs, it was important that their modified size after scanning and their resolution (pixels to the inch) were matched, and that in turn their resolution matched that of the Taj Mahal background. Similarly, in placing them against a black background, the background was sized to allow a small margin all round, once again at the same resolution.

Satisfied with the eclipse sequence, I then saved the image and clicked with the magic wand on the uniform black background. This selected all of the 'sky' and a further click on the inverse button changed the selection to the solar images linked as a group. The Taj Mahal image was then called up. The move tool was activated and the mouse used to drag the solar images across to the Taj Mahal image. In Photoshop this then became a two-layered image. With layering, you activate a layer by clicking on its title bar in the layers menu and any work you do affects that layer only. The sequence of solar images was positioned carefully to my satisfaction in the sky and then I flattened the image—i.e. converted it from two layers to one with the solar images now forming an integral part of a single picture. (Often one leaves this step until a composite is totally finished but since the finished product was not going to be very complex I saw no reason to delay the step.)

One thing remained. Enough of the top of the palace was reflected in the water in front for reflections of one or two of the partial phases to have been reflected, too, if the eclipse had actually taken place as portrayed. To do this, I reverted to the solar sequence on the black background and selected the two crescents that I calculated would have been reflected. These were then moved across to the composite. Positioning them with the move tool was no problem but of course their angle was wrong to serve as reflections. Here I needed to use the free transform function in Photoshop's Edit menu which enabled me in effect to invert and laterally reverse the two crescents which existed on their own layer until I again flattened the image. The picture was complete.

The impression of how the 11 August eclipse might well appear at Stonehenge was constructed in essentially the same way (figure 19). The original picture was taken at Stonehenge at dawn looking towards the south east some years ago—the sky appearing not unlike that during an eclipse. Stars and planets not wanted in the eclipse impression were removed from the original. An estimate was made of the extent of the photosphere that will be seen at Stonehenge when the eclipse is at its maximum there—over 98% —and an appropriate disc was created and dropped into the mid-point of an existing eclipse sequence image. This was then imported into the sky of the modified Stonehenge image over the standing stones.

The image appearing on the cover is markedly less fanciful and is intended to show as much detail as possible of the corona during an eclipse. The problem with photographing the corona is its very considerable contrast range between the bright inner region and the relatively dim outer corona. By altering exposures, inner and outer corona can be well recorded in a sequence of individual images but they cannot be imaged together in a single image. The addition of a radial gradient filter—which absorbs light from the inner corona more strongly than that from the outer—to the front of a lens overcomes the problem. But such a filter is expensive and for maximum effect must be dedicated to use with one lens. Another way of ending up with a full range of tones in the entire corona is for very skilled printing of a series of images on to a single sheet of photographic paper in the darkroom. This calls for great talent and is extremely time consuming.

The personal computer has now made additional ways possible and in the January 1998 issue of *Sky & Telescope* an article by Gerald Pellett, a scientist at NASA's Langley Research Centre in the US, outlined a method using Photoshop. A number of different exposures of the corona are scanned and each image is matched exactly as regards area, position of the lunar disc and so on to its fellows—a most important requirement because the images will be superimposed on each other and, whether this is done by digital or analogue methods, great precision is required to achieve an optimum effect. Each image is then processed using Photoshop's radial blur filter with the 10° spin mode selected. Using the programme's Apply Image function, the pixel brightness values of the blurred image are subtracted from those in the original, which results in a *difference image* with a small range of pixel values. When working with colour transparency originals, this image appears as a light silver grey picture on the monitor with considerable detail of the corona already showing. As the work proceeds, the difference images are added sequentially one to another with the final, sum image being multiplied by the pixel values of one of the original pictures.

The picture shown here was created from a six original transparencies shot of the corona on ISO200 film in Guadeloupe during the 26 February 1998 total eclipse. The exposures ranged from 1/8 to 4 seconds and the lens used was a 1000 mm f/11 mirror reflex. As intended, the picture shows infinitely more detail than that contained in any of the individual originals. But the technique involved is obviously one which is capable of further development with the aim of improving the detail seen in the corona and from my experience seems to be limited primarily by the amount of time available for experimentation.

Captions to Figures

Figure 1. A GOES weather satellite image showing the Moon's shadow over NW South America and the Caribbean during the solar eclipse of 26 February 1998. (Courtesy US National Oceanic and Atmospheric Administration.)

Figure 2. Computer simulation of the sky at mid-totality on 11 August 1999, as seen from Redruth, Cornwall.

Figure 3. The atmosphere of an eclipse. From top left clockwise: an offering of eclipse rum in Guadeloupe, 26 February 1998 and a poster displayed on the island; a solar eclipse in the Philippines in March 1988 was celebrated with a stamp issue; on eclipse trips expert advice is always available, even if the well known expert in this case was not in his normal attire; a celebratory drink after a magnificent eclipse seen from La Paz, Mexico in July 1991; photographers get to work during the annular eclipse visible from Arizona in May 1994; during the 1988 Philippines eclipse, locals gather around the visitors' cameras.

Figure 4. (*a*) The size of the Sun as it appears in the entire 35 mm negative area of a 'fun' camera and (*b*) a straight fun camera picture of an evening twilight sky with an actual eclipsed Sun superimposed.

Figure 5. The Fuji 100 mm throwaway camera yields a much bigger Sun. The main pictures here show the size of the Sun (appearing a light blue because of the use of a Mylar filter) at top and with a Wratten 21 (orange) filter added at bottom. The 'atmospheric' images inset result from the use of the camera's slower 'Flood light' setting.

Figure 6. The size of the Sun recorded using different focal length lenses: (top to bottom) 50 mm, 180 mm, 300 mm, 450 mm and 1000 mm.

Figure 7. (*a*) A time-lapse test run with 50 mm lens over the period occupied by the entire eclipse, with one exposure every ten minutes: inset is the result of exposing every five minutes. Below (*b*) is the result using a 180 mm focal length lens.

Figure 8. A reserve camera is a good idea; in this picture taken during the eclipse of 18 March 1988, a hand-held Nikon camera with a 180 mm f/2.8 focal length lens fitted satisfactorily recorded both prominences and the inner corona during totality.

Figure 9. Eclipse atmosphere recorded using ordinary cameras; during the July 1991 eclipse seen from Mexico, totality is near (top) in this picture looking out over the Pacific towards the onrushing lunar shadow and below, with street lights clearly seen, totality has arrived but the western horizon is already clear of the Moon's shadow. Bottom are multiple crescent Suns, projected by the leaf canopy of a tree.

Figure 10. Mood pictures of the Sun in eclipse: (*a*) an Egyptian vulture comes to roost as the Moon's umbra passes over India on 24 October 1995 and (*b*) a hot air balloon forms an attractive composition with the crescent Sun and a solar glitter path over the Pacific during the annular eclipse seen from California on 4 January 1992.

Figure 11. Annular eclipse of the Sun, Arizona, 10 May 1994.

Figure 12. A partial eclipse of the Sun was seen from the UK on 12 October 1996. Compare the lack of sunspots on the disc at this time with the number shown in the pictures taken at a time of near maximum (page 47, figure 17).

Figure 13. The Diamond Ring and a single Bailys' Bead are seen together with the inner corona in this picture taken in Guadeloupe in February 1998.

Figure 14. A magnificent solar prominence was seen during the lengthy eclipse of 11 July 1991.

Figure 15. Limb detail: Baily's Beads, the chromosphere and prominences are seen clearly in this picture taken from the Atacama Desert in Chile on 3 November 1994 (*a*). A longer exposure records the developing corona with the remnants of the chromosphere adding a delightful reddish tone (*b*).

Figure 16. A beautiful time lapse image recorded by Akira Fujii and other Japanese colleagues during the 11 July 1991 eclipse. The cathedral in central La Paz provides the foreground for a sequence of images taken at 5 minute intervals using a Mamiya Press camera with a 100 mm focal length lens fitted. 64ISO Kodachrome film was used and the exposure at totality was $\frac{1}{4}$ second at f/4. Notice the elliptical track of the Sun, which can also be seen in the 50 mm test image appearing on page 44 (figure 7(*a*).

Figures 17, 18 and 19. These are described fully by the text of chapter 7.

In the interests of brevity, the photographic detail given in the picture captions is limited. Generally, however, exposures were based on the table given on page 24, using ISO100 colour transparency film loaded in Nikon F3 or F4S cameras fitted (unless otherwise indicated) with a 1000 mm f/11 Nikkor mirror-reflex lens.

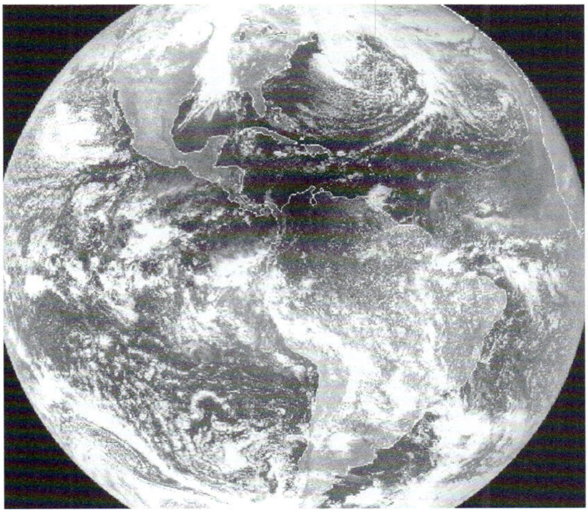
1.

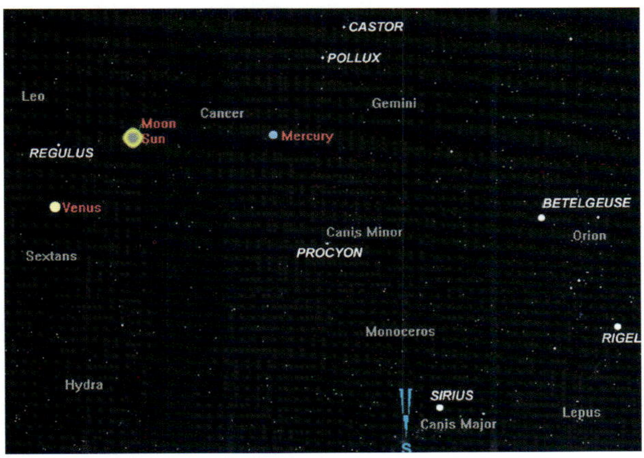
CASTOR
POLLUX
Leo
Cancer
Gemini
Moon
Sun
Mercury
REGULUS
Venus
BETELGEUSE
Canis Minor
Orion
Sextans
PROCYON
Monoceros
RIGEL
Hydra
SIRIUS
Canis Major
Lepus
S
2.

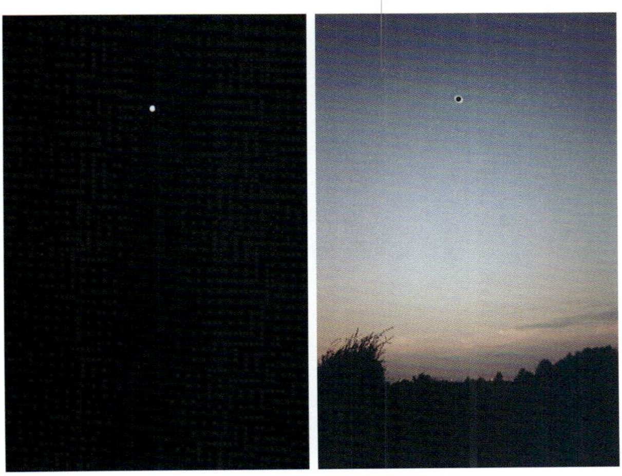
4.

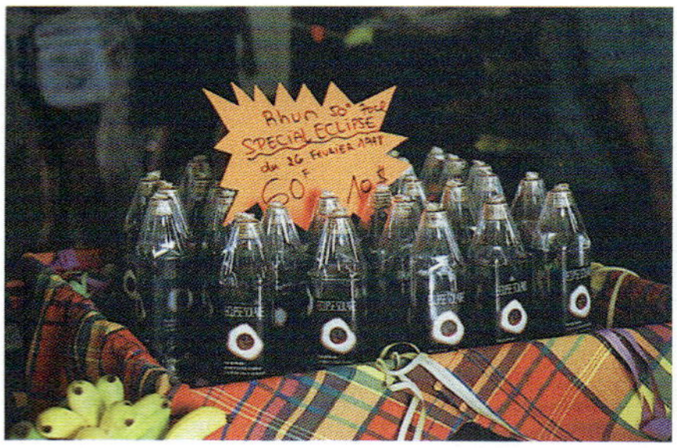
SPECIAL ECLIPSE

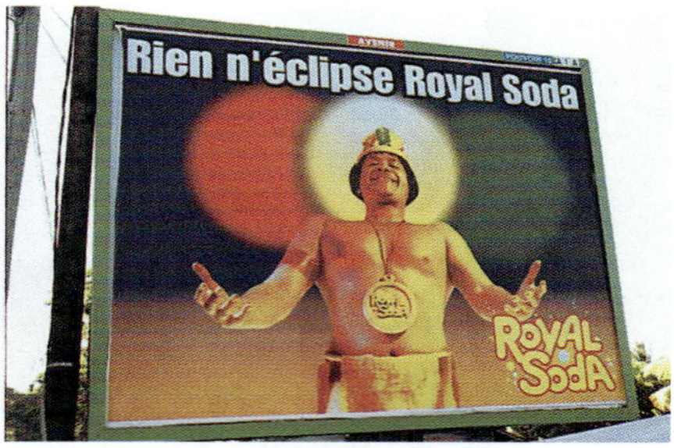
Rien n'éclipse Royal Soda
ROYAL SODA

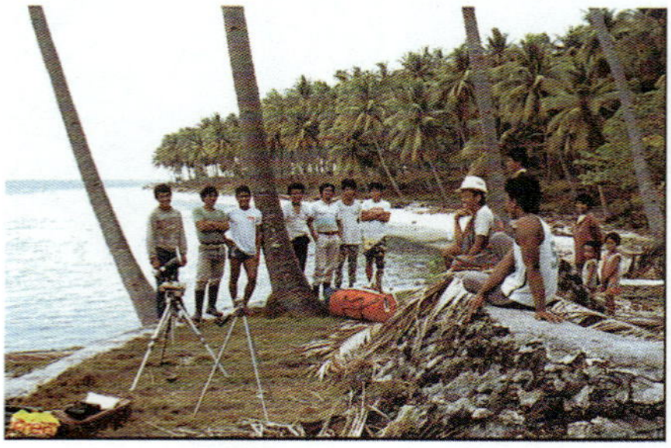

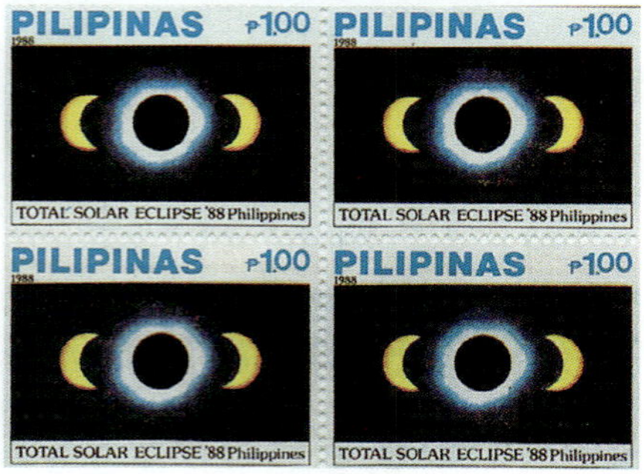
PILIPINAS P1.00
1988
TOTAL SOLAR ECLIPSE '88 Philippines
PILIPINAS P1.00
1988
TOTAL SOLAR ECLIPSE '88 Philippines
PILIPINAS P1.00
1988
TOTAL SOLAR ECLIPSE '88 Philippines
PILIPINAS P1.00
1988
TOTAL SOLAR ECLIPSE '88 Philippines

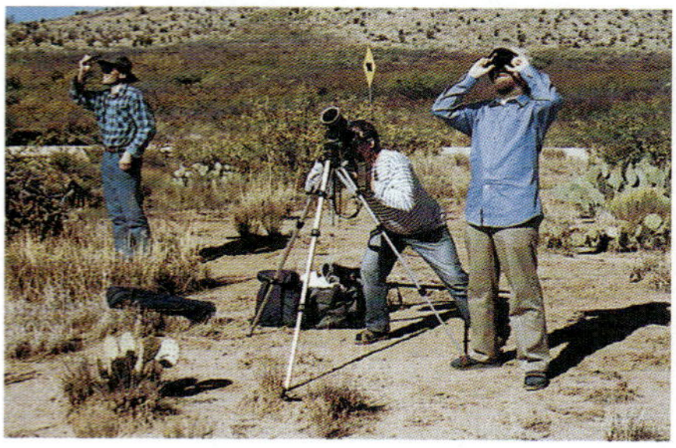

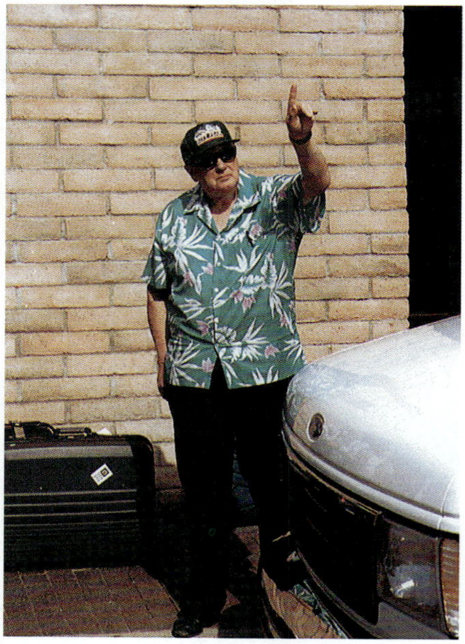

3.

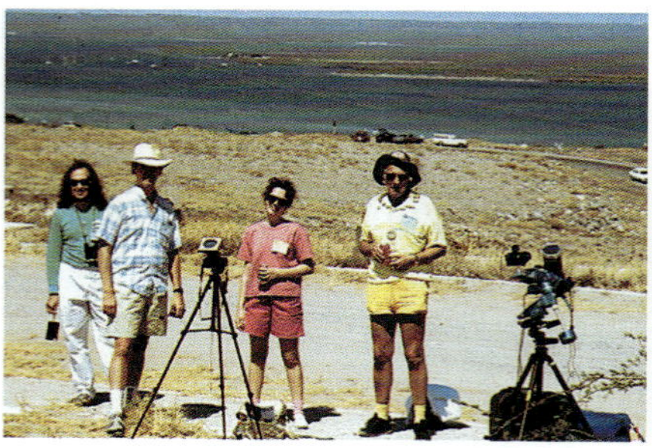

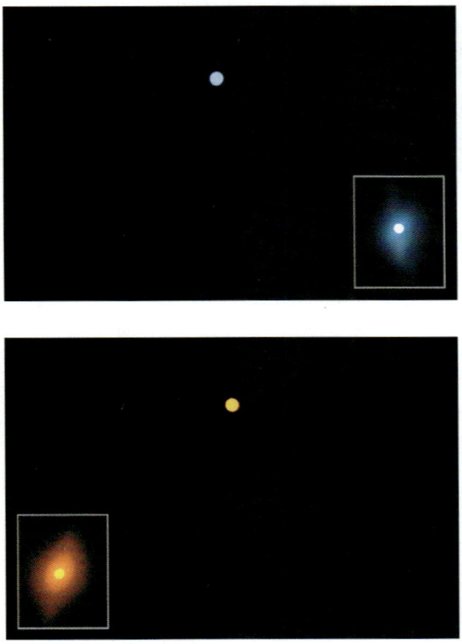

5.

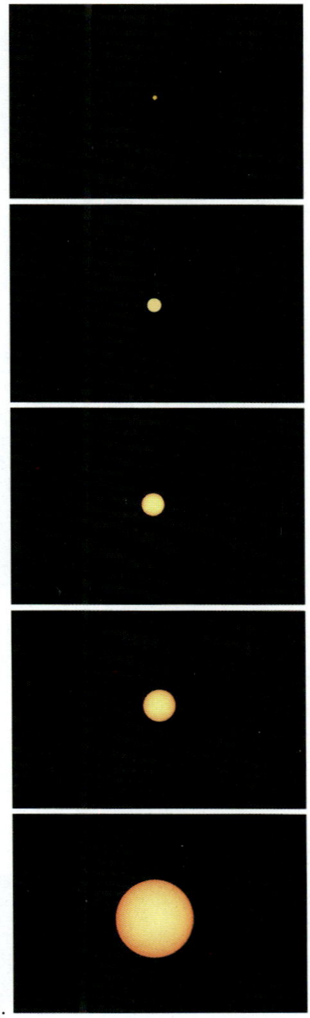

6.

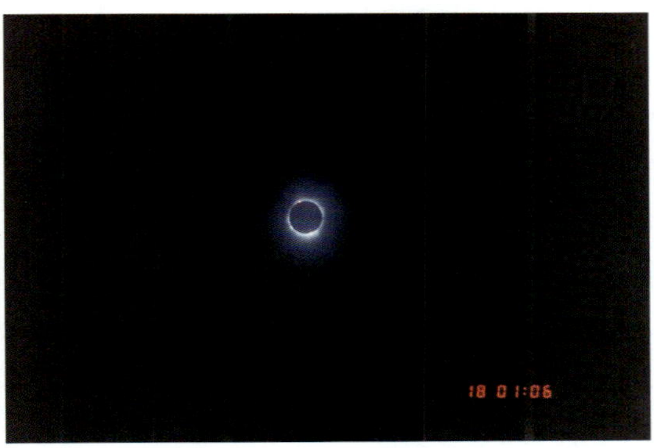

8.

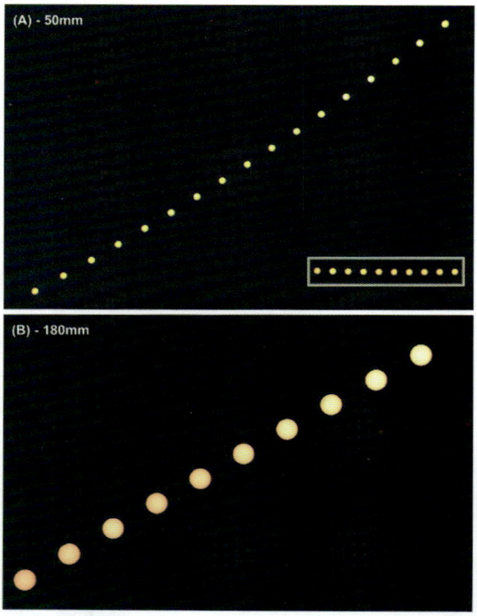

7.

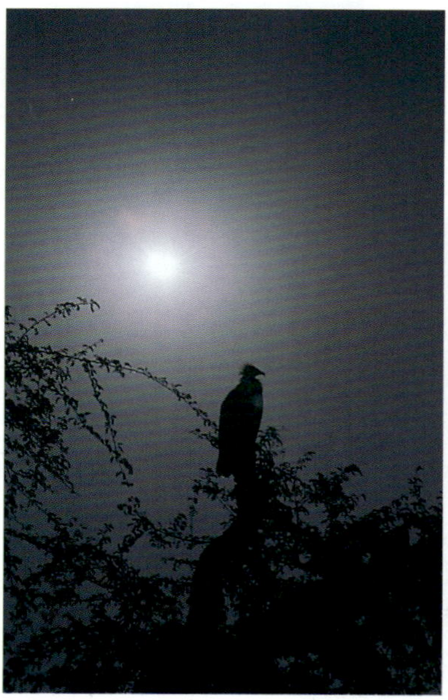
10(a).

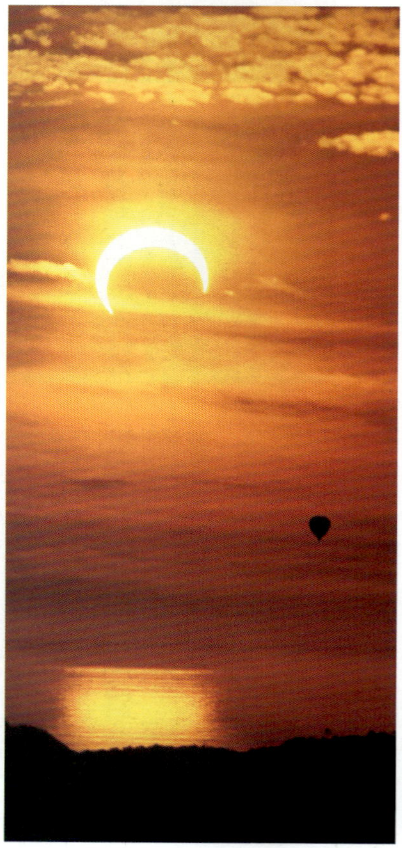
10(b).

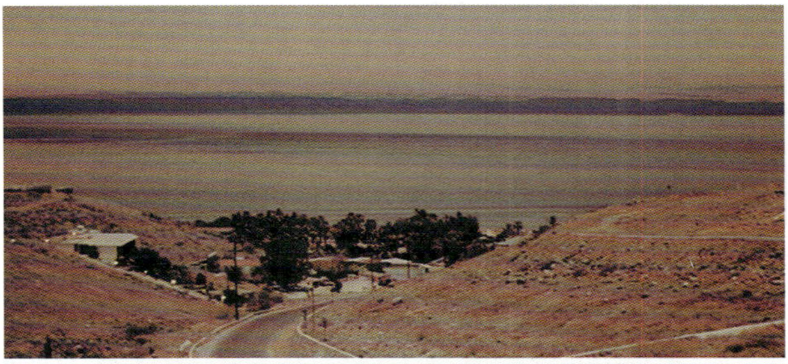

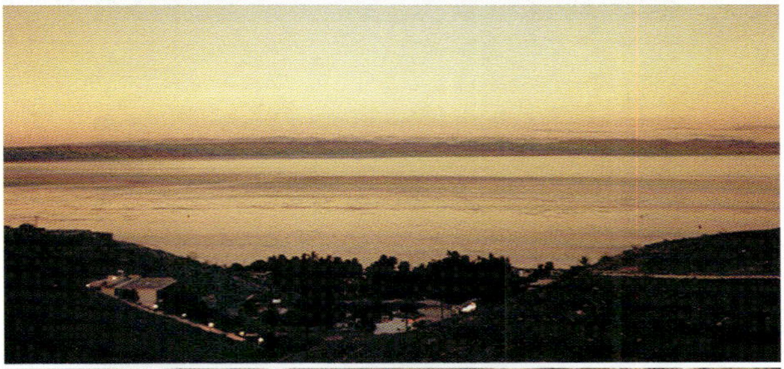

9. 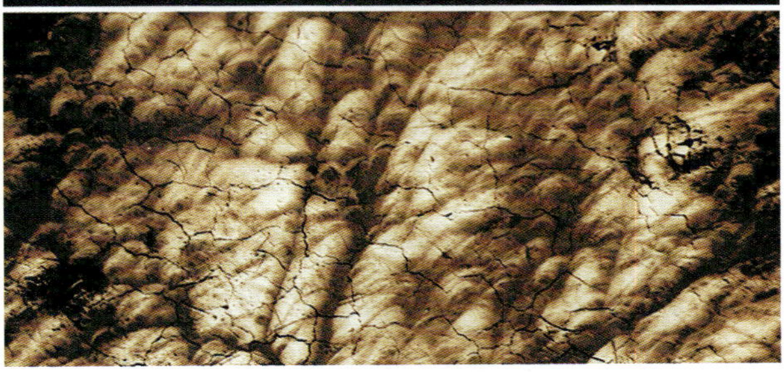

11. 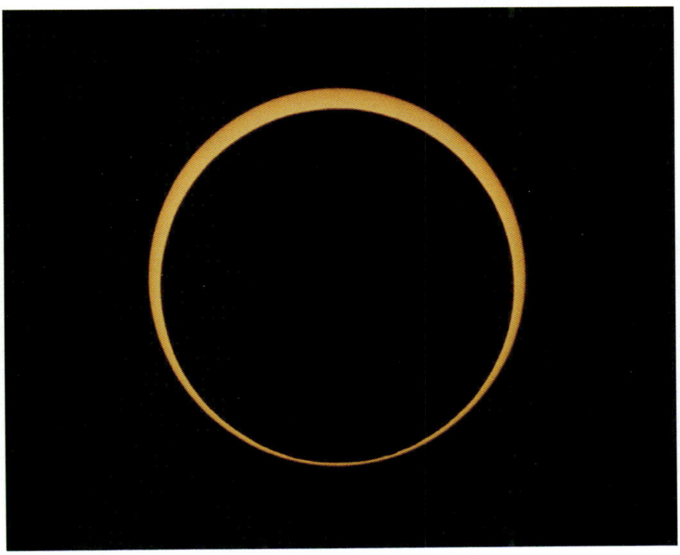

12. 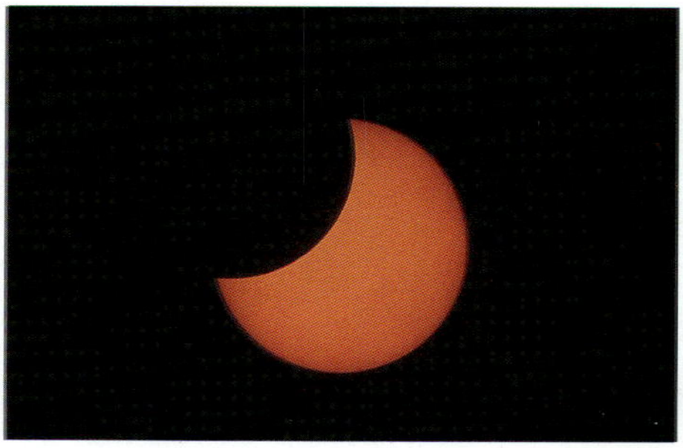

13. 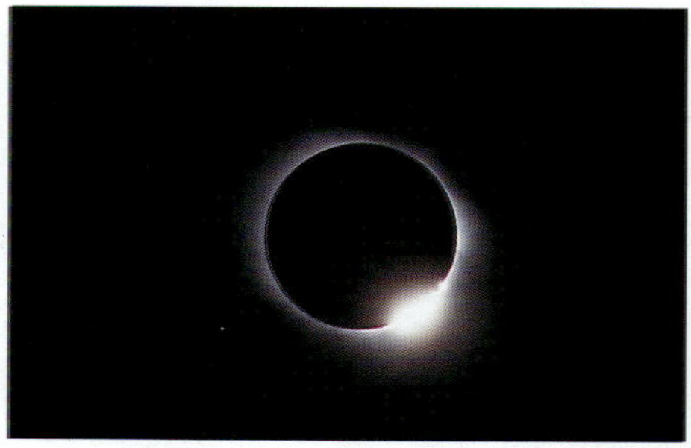

14. 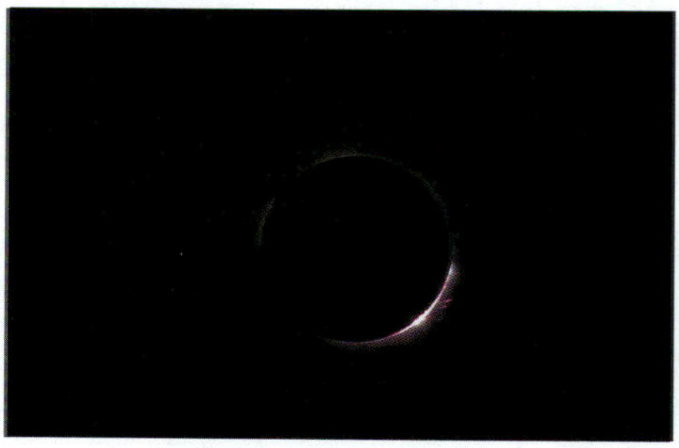

15(a). 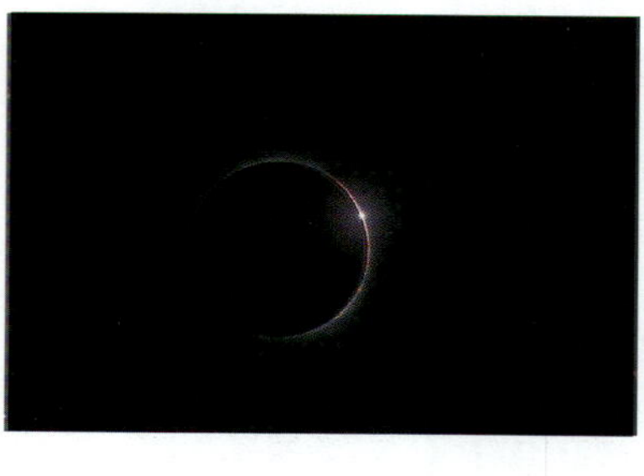

15(b). 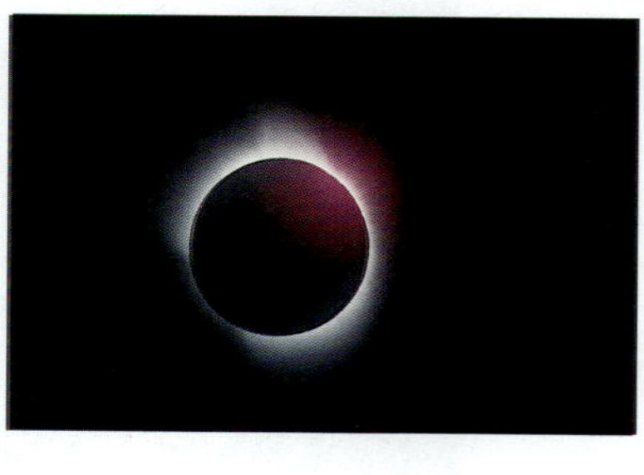

16. 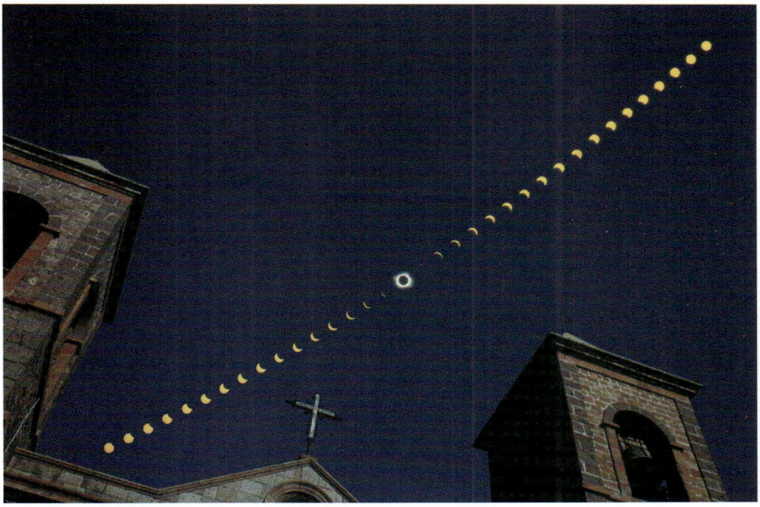

17. 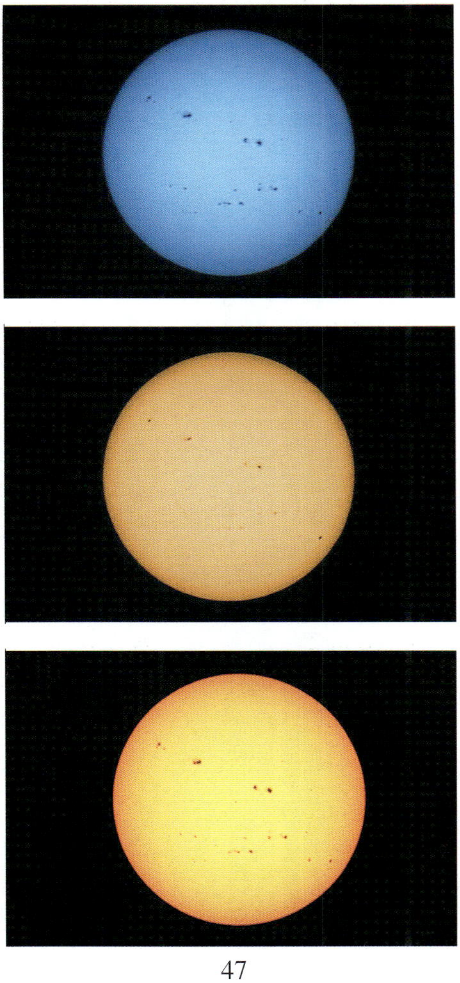

18. 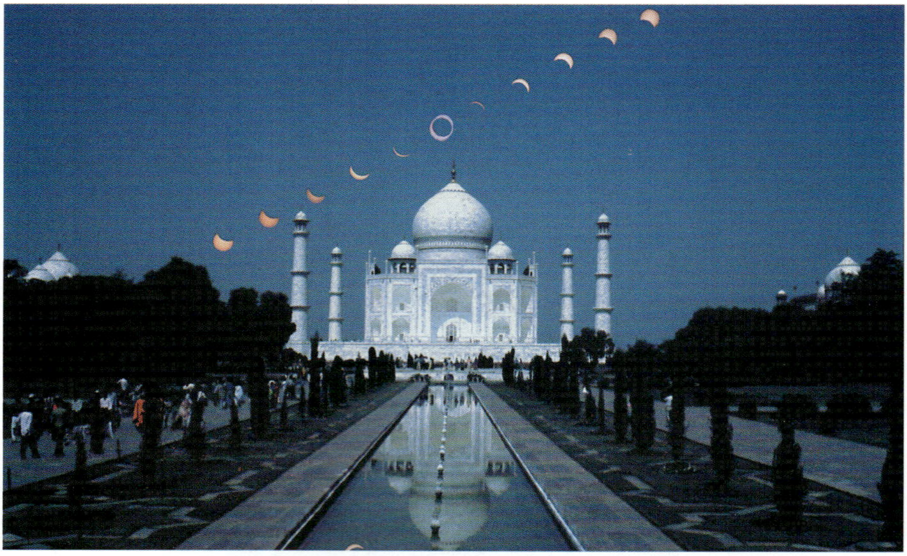

19. 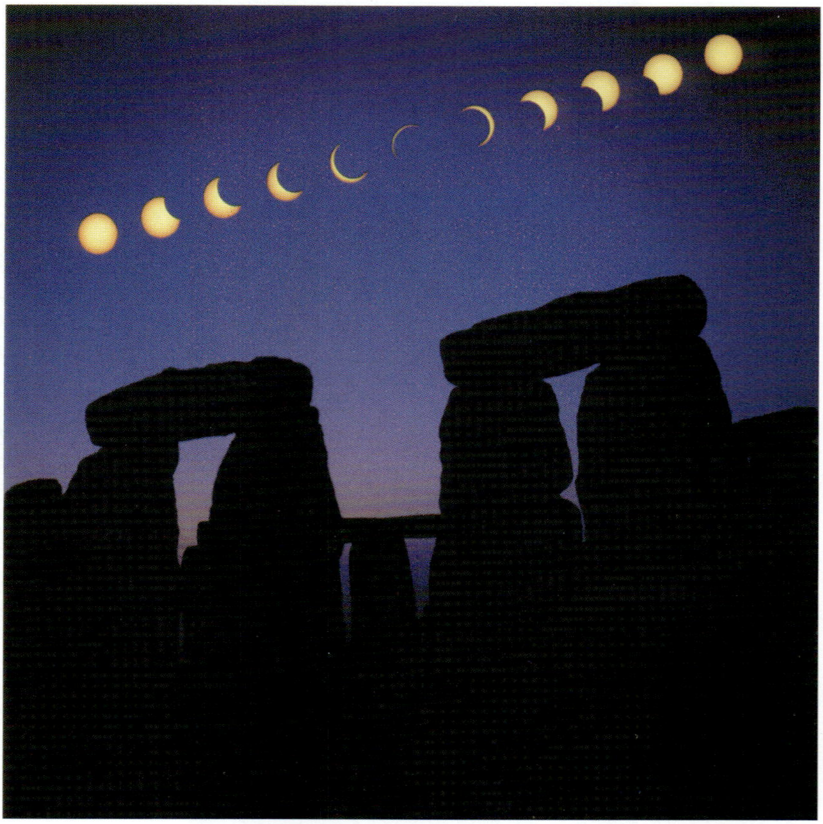

Chapter 8

A Foreign Eclipse?

This primer is directed principally to readers who plan to see or photograph the eclipse in England. Chapter 2, however, did point out that the greatest extent of the eclipse and statistically the best chances of clear skies would occur far to the south east—in Rumania and Turkey, if we omit such politically sensitive spots of the world as Syria, Iraq and Iran. The prospect of seeing an eclipse at home will doubtless entice even some inveterate eclipse chasers to try their luck in Cornwall and Devon but others amongst this book's intended readership may feel more exotically inclined. Brief additional photographic guidance is thus in order.

The points made in the two earlier chapters all apply: good photographic practice, tests well in advance of eclipse day, leaving as little to chance as possible are as important if not more so when you leave the UK. Some aspects need to be emphasized.

- A tripod might be considered a nuisance on foreign travel but if you have the more advanced type of camera that can use one, take it. Bear in mind, too, that if Rumania is your destination, the Sun will be at its greatest elevation there (almost 60°) so your pre-journey tests should have checked that there are no problems with the tripod bearing the weight of the camera/lens combination at that steep angle and of being locked at that position.
- Take spare batteries if the camera uses them. Also take plenty of your chosen film(s). Do not rely on getting them where you are going; moreover, film bought at home is likely to have been stored in far better conditions than in many outlets abroad.
- If hiring a car when abroad, keep your photographic equipment and films in as cool a location as possible—certainly not on the back window sill!
- Whilst on the subject of heat in Rumania or Turkey in August, remember to take a light cloth with you to drape over the camera/tripod during the eclipse. Heat never does equipment or films any good and touching the metal of tripod or camera during mid-day temperatures will not be very pleasant. You should take the normal personal precautions on sudden exposure to high temperatures and the Sun; also, you are likely to be using plenty of sun cream and it will be difficult to keep it away from camera surfaces and the front element of your lenses. Take plenty of wipes therefore and—better still so far as lenses are concerned—take lens cleaning wipes or cleaning solution with you.
- Camera jams can occur and I always find a changing-bag useful on such occasions because you can normally retrieve a film safely without spoiling any (or at least many) frames. A set of jeweller's screwdrivers is also valuable for tightening small screws used on cameras, filters and other accessories. If you are making *ad hoc* arrangements for fitting filters and other items, do not forget to take spare tape: you are very unlikely to find any readily available where you are going. If you have an advanced camera, take its user's guide with you—I am

always surprised by how much one can forget about a camera's operation, safety procedures and other information.

- If you have expensive equipment, remember that travel insurance cover can impose very low upper value limits on individual items. However, you may find that if the equipment is covered under your household insurance a worldwide extension for up to 30 days applies. Check with your insurer.

Film needs some further guidance. Photographic film can be affected by exposure to x-rays and this may cause you some worry, particularly as the procedures followed at UK airports do not give you the choice of asking for a hand search. Have no fears about films loaded in your hand luggage and passed through the scanners at British airports. Tests have shown the machines to be perfectly safe—*but on no account put your films in luggage which is going into the aircraft hold.* Abroad, I have rarely found any problem with requesting security staff to hand check films loaded in a clear plastic bag, particularly when arriving well in advance of flight departure.

On the assumption of a successful eclipse, you will have to decide about having your (hopefully) hard-won images processed. Getting them processed abroad removes any further worries about possible x-ray damage on the return journey. However, I have always preferred either to process film myself on the spot or, if this is not possible for any reason, to bring it back to laboratories at home that I know work to the highest standards. The latter requires some patience but is worthwhile, I think, because I would be greatly concerned about having my negative colour film processed, cut and printed by an establishment, dealing with masses of the usual holiday films, about whose quality control I knew nothing.

There is one further piece of advice for the owners of advanced and therefore most likely very expensive cameras and other items—particularly if quite new. Take the receipts with you on the trip. Very probably you will not have problems going (although some countries seem to be concerned about the possible resale of expensive equipment brought in by tourists) but you may when returning home. However, it is easy enough to prevent difficulties from occurring.

Considerable detail about the advantages that different countries offer for seeing the 11 August 1999 eclipse appears in the World Wide Web sources that appear in appendix D. Readers who feel tempted to go abroad have two broad possibilities before them. Experienced travellers may decide to make their own arrangements for an independent expedition. A major problem will be accommodation. Hotels are booked up years in advance for the period of an eclipse and, unless the visitor has friends or acquaintances who can help, securing accommodation may be impossible. This will not be the case with the second possibility—joining one of the trips organized by a specialist astronomical tour agency (see appendix F). Such agencies will have planned visits of different durations years in advance, will have already selected a good site for viewing the eclipse (as well as offering other attractions), and above all can offer assured accommodation—although even with them, double booking at prices to the distinct advantage of the hoteliers is not exactly unknown! The comforting thing in such cases is that it is the tour organizers' headache, not yours.

The inexperienced eclipse chaser, whether photographically knowledgeable or not, will have the further advantage on such a specialist trip of help and guidance from experienced amateur astronomers and eclipse chasers who are likely to make up the biggest element of the party. By and large, they are keen to share their enthusiasm and advice—indeed, the problem may be a little too much enthusiasm on their part which will tend to overwhelm the uninitiated. But there is no doubting the good intentions and frequently the very considerable knowledge that will give the eclipse far greater meaning for those who might be seeing one for the first time.

Postscript

I very much hope that the advice given in the previous pages will help many people secure good images of some aspect of the solar eclipse on 11 August 1999—weather permitting.

I would like to end by telling two stories, each with basically the same moral. I saw the eclipse of 11 July 1991 from La Paz, in Baja California, Mexico. Totality was over 6 minutes 50 seconds long and was magnificent. Next to me during the eclipse was a young American. It was his first eclipse but he had come armed with a very detailed photographic programme that he was determined to conduct. Totality began and virtually non-stop during the almost seven minutes that it lasted, all I could hear from him were exclamations of amazement, delight and awe. He took very, very few pictures but he positively revelled in the glories of the eclipse. The moral? If you are photographing a solar eclipse—keep it simple and most of all enjoy the spectacle.

As to the second story: it has been my privilege over the years to see many rocket launches from the Kennedy Space Centre in Florida, in the same way as I have been privileged to witness numerous eclipses. However, for me, photographic coverage has always been a prime requirement during both forms of spectacle. When you are conducting such work, you can eventually enjoy seeing what will (hopefully) be an excellent record of the events—but you cannot really see or experience them at the time because you are concentrating on a host of details. I keep promising myself that one of these days I will just watch a launch or, more importantly, just watch a solar eclipse and delight in the spectacle.

I have not managed that yet but one day I might. Perhaps on 11 August 1999?

APPENDIX A

ECLIPSE TIMES FOR TOWNS EXPERIENCING TOTALITY .

TOWN	FIRST CONTACT	Alt (°)	SECOND CONTACT	THIRD CONTACT	FOURTH CONTACT	Alt (°)	MAXIMUM ECLIPSE	ALT (°)	AZI (°)	UMBRAL DURATION
	h. m. s.		h. m. s.	h. m. s.	h. m. s.		h. m. s.			m. s.
ALDERNEY	09 59 39	38	11 15 20	11 17 01	12 37 15	55	11 16 11	48	135	01 41
BODMIN	09 57 50	36	11 12 22	11 13 40	12 32 59	53	11 13 01	46	131	01 19
CAMBORNE	09 56 58	35	11 11 01	11 13 02	12 32 03	53	11 12 02	46	130	02 01
FALMOUTH	09 57 08	36	11 11 19	11 13 21	12 32 28	54	11 12 20	46	130	02 02
HELSTON	09 56 53	36	11 11 01	11 13 02	12 32 08	54	11 12 02	46	130	02 01
HUGH TOWN	09 55 45	35	11 09 42	11 11 23	12 30 29	53	11 10 32	45	128	01 41
KINGSBRIDGE	09 58 34	37	11 13 17	11 15 09	12 34 33	54	11 14 13	47	133	01 52
NEWQUAY	09 57 25	36	11 11 40	11 13 18	12 32 24	53	11 12 29	46	131	01 38
NEWTON ABBOT	09 58 59	37	11 14 20	11 14 50	12 34 47	54	11 14 35	47	133	00 30
PENZANCE	09 56 39	35	11 10 39	11 12 41	12 31 41	53	11 11 40	46	130	02 02
PLYMOUTH	09 58 16	36	11 12 54	11 14 32	12 33 53	54	11 13 43	46	132	01 39
REDRUTH	09 57 03	36	11 11 09	11 13 10	12 32 11	53	11 12 09	46	130	02 01
SAINT AUSTELL	09 57 36	36	11 11 54	11 13 44	12 32 52	53	11 12 49	46	131	01 50
SALTASH	09 58 15	36	11 12 53	11 14 28	12 33 50	54	11 13 41	46	132	01 35
TORQUAY	09 59 01	37	11 14 08	11 15 15	12 34 57	54	11 14 41	47	133	01 07
TRURO	09 57 17	36	11 11 27	11 13 25	12 32 29	53	11 12 26	46	130	01 57

(KEY: **ALT** is altitude of Sun in degrees at indicated time. **AZI** is azimuth or direction of the Sun at indicated time – e.g. 90° is due East and 180° is due South. All times are British Summer Time.)

APPENDIX B.

ECLIPSE TIMES FOR TOWNS EXPERIENCING PARTIAL ECLIPSE.

TOWN	FIRST CONTACT	ALT (°)	MAXIMUM ECLIPSE	ALT (°)	AZI (°)	MAXIMUM EXTENT	FOURTH CONTACT	ALT (°)
ENGLAND	h. m. s.		h. m. s.				h. m. s.	
BIRMINGHAM	10 02 52	37	11 17 59	46	138	0.935	12 37 00	52
BLACKBURN	10 03 40	37	11 17 45	45	138	0.892	12 35 35	51
BLACKPOOL	10 03 13	36	11 17 05	45	137	0.890	12 34 45	51
BOLTON	10 03 31	37	11 17 44	45	138	0.898	12 35 43	51
BOURNEMOUTH	10 00 55	38	11 17 04	48	136	0.995	12 37 29	54
BRADFORD	10 04 26	37	11 18 43	46	139	0.889	12 36 37	51
BRIGHTON	10 02 57	39	11 19 39	49	139	0.988	12 40 16	54
BRISTOL	10 00 58	37	11 16 27	47	136	0.972	12 36 11	53
CHESTEFIELD	10 04 07	37	11 18 53	46	139	0.908	12 37 19	52
COVENTRY	10 03 07	38	11 18 23	47	138	0.937	12 37 32	52
DERBY	10 03 41	38	11 18 39	46	139	0.919	12 37 22	52
DUDLEY	10 02 37	37	11 17 39	46	137	0.935	12 36 37	52
EXETER	09 59 16	37	11 14 48	47	133	0.997	12 34 53	53
GLOUCESTER	10 01 47	37	11 17 09	47	137	0.957	12 36 36	53
HIGH WYCOMBE	10 03 03	38	11 19 03	48	139	0.962	12 38 57	53
HUDDERSFIELD	10 04 13	37	11 18 36	46	139	0.894	12 36 38	51
IPSWICH	10 05 34	40	11 21 56	48	143	0.942	12 41 44	53
KINGSTON (HULL)	10 05 47	38	11 20 32	46	141	0.888	12 38 40	51
LEEDS	10 04 38	37	11 18 57	46	139	0.888	12 36 50	51
LEICESTER	10 03 46	38	11 19 03	47	139	0.928	12 38 04	52
LIVERPOOL	10 02 51	36	11 17 01	45	137	0.905	12 35 05	51
LONDON	10 03 33	39	11 19 51	48	140	0.966	12 39 57	53
LUTON	10 03 40	39	11 19 39	48	140	0.953	12 39 24	53
MANCHESTER	10 03 36	37	11 17 55	46	138	0.901	12 36 01	51
MANSFIELD	10 04 15	38	11 19 09	46	139	0.911	12 37 42	52
MIDDLESBRO'	10 05 53	37	11 19 45	45	141	0.861	12 37 00	50
NEWCASTLE	10 06 02	37	11 19 30	45	140	0.848	12 36 20	50
NORTHAMPTON	10 03 32	38	11 19 07	47	139	0.942	12 38 31	53
NORWICH	10 06 18	39	11 22 18	48	143	0.922	12 41 38	53
NOTTINGHAM	10 04 03	38	11 19 05	47	139	0.917	12 37 49	52
OXFORD	10 02 40	38	11 18 26	47	138	0.959	12 38 09	53
POOLE	10 00 50	38	11 16 57	48	136	0.995	12 37 21	54
PORTSMOUTH	10 01 52	38	11 18 16	48	138	0.991	12 38 46	54
PRESTON	10 03 29	36	11 17 29	45	137	0.892	12 35 16	51
READING	10 02 39	38	11 18 41	48	138	0.969	12 38 41	53
SHEFFIELD	10 04 12	37	11 18 50	46	139	0.903	12 37 09	51
SOUTHAMPTON	10 01 38	38	11 17 50	48	137	0.988	12 38 11	54
SOUTHEND	10 04 34	39	11 21 07	49	141	0.961	12 41 18	54
STOKE-ON-TRENT	10 03 06	37	11 17 47	46	138	0.918	12 36 19	52
SUNDERLAND	10 06 09	37	11 19 43	45	141	0.850	12 36 39	50
SWINDON	10 01 55	38	11 17 37	47	137	0.967	12 37 24	53
WALSALL	10 02 49	37	11 17 50	46	138	0.932	12 36 46	52
WATFORD	10 03 27	39	11 19 34	48	140	0.960	12 39 29	53
WEST BROMWICH	10 02 47	37	11 17 51	46	138	0.934	12 36 50	52
WOLV'RHAMPTON	10 02 40	37	11 17 38	46	137	0.932	12 36 31	52
YORK	10 05 17	37	11 19 39	46	140	0.882	12 37 30	51

SCOTLAND							
ABERDEEN	10 08 27	36	11 20 04	43 141	0.776	12 34 41	48
AYR	10 03 53	35	11 16 08	43 136	0.837	12 32 02	49
DUNDEE	10 06 40	35	11 18 37	43 139	0.801	12 33 49	48
DUNFERMLINE	10 05 42	35	11 17 50	43 138	0.815	12 33 21	48
EDINBURGH	10 05 47	35	11 18 04	43 138	0.818	12 33 45	49
FALKIRK	10 05 20	35	11 17 25	43 138	0.818	12 32 58	48
GLASGOW	10 04 47	35	11 16 50	43 137	0.822	12 32 25	48
GREENOCK	10 04 27	35	11 16 18	43 136	0.821	12 31 44	48
KIRKCALDY	10 06 03	35	11 18 13	43 139	0.813	12 33 45	48
WALES							
CARDIFF	10 00 22	37	11 15 36	46 135	0.972	12 35 11	53
NEWPORT	10 00 41	37	11 15 57	46 135	0.969	12 35 30	53
PORT TALBOT	09 59 56	36	11 14 54	46 134	0.969	12 34 17	53
SWANSEA	09 59 49	36	11 14 42	46 134	0.968	12 34 01	53
IRELAND (NORTH)							
BELFAST	10 01 37	34	11 14 03	43 133	0.868	12 30 31	49
IRELAND							
CORK	09 56 03	33	11 09 06	43 127	0.963	12 27 06	51
DUBLIN	09 59 42	34	11 12 49	43 131	0.912	12 30 16	50
CHANNEL ISLANDS							
ST PETER PORT	09 59 03	38	11 15 34	48 134	0.999	12 36 45	55

(KEY: **ALT** is altitude of Sun in degrees at indicated time. **AZI** is azimuth or direction of the Sun at those times – e.g. 90° is due East and 180° is due South. **Maximum Extent** is an indication of the amount of the Sun obscured by the Moon: 1.000 would be total. All times are British Summer Time.)

Appendix C: Further Reading

Readers wishing to pursue information about the eclipse of 11 August 1999 and solar eclipses generally will find the following books published in the UK of use:

The RGO Guide to the 1999 Total Eclipse of the Sun by Steve Bell (Royal Greenwich Observatory, 1996); ISBN 0-905-087-03-8.

The Sun in Eclipse by Michael Maunder and Patrick Moore (London: Springer, 1998); ISBN 3-540-76146-2.

UK Solar Eclipses from Year 1 by Sheridan Williams (Clock Tower Press, 1996); ISBN 1-85142-093-2.

Astro-photography: An Introduction by H J P Arnold (George Philip, 1998 (Reprinted)); ISBN 0-540-06086-0.
(Recommended for amateur photographers who develop an interest in pursuing astro-photography generally.)

Appendix D: Information on the World Wide Web

The address of the Solar Eclipse 1999 UK Co-ordinating Group is: *http://www.eclipse.org.uk* This has numerous links to members of the Group—such as the Royal Astronomical Society and the British Astronomical Association—as well as detailed information about accommodation available in Cornwall.

NASA Reference Publication 1398 by Fred Espenak and Jay Anderson—Total Eclipse of 11 August 1999 has already been mentioned. The full text of that publication plus updating material is available from: *http://sunearth.gsfc.nasa.gov/eclipse/TSE1999/TSE1999.html*

Fred Espenak's eclipse home page is also well worth visiting with innumerable links to other sites: *http://sunearth/gsfc.nasa.gov/eclipse/eclipse.html*

Appendix E: Suppliers

The author has had direct dealings in the recent past with the following suppliers to his complete satisfaction:

Broadhurst Clarkson & Fuller Ltd, Telescope House, 63 Farringdon Road, London EC1M 3JB. Tel: 0171 405 2156. Fax: 0171 430 2471.

David Hinds Ltd, Unit 34, The Silk Mill, Brook Street, Tring, Herts HP23 5EF. Tel: 01442 827768. Fax: 01442 890763.

Venturescope, The Wren Centre, Westbourne Road, Emsworth, Hampshire PO10 7RN. Tel: 01243 379322. Fax: 01243 370717.

Baader Planetarium, Zur Sternwarte, 82291 Mammendorf, Germany. Tel: 0049 8145 8802. Fax: 0049 8145 8805.

Eclipse99 Limited, Belle Etoile, Rue du Hamel, Castel, Guernsey GY5 7QJ. Tel: 01481 64847. Fax: 01481 64871.

Details of other astronomical suppliers will be found in the astronomical press.

Appendix F: Astronomical Tours

Journeys of Special Scientific Interest, 217 Monton Road, Monton, Manchester M30 9PN. Tel: 0161 707 8547. Fax: 0161 707 9190.

Explorers Tours, 223 Coppermill Road, Wraysbury TW19 5NW. Tel: 01753 680237. Fax: 01753 682660.